CONCEPTS IN THERMAL COMFORT

CONCEPTS IN THERMAL COMFORT

M. DAVID EGAN
Associate Professor of Architecture
Clemson University

PRENTICE-HALL, INC., *Englewood Cliffs, New Jersey*

Library of Congress Cataloging in Publication Data

Egan, M David.
    Concepts in thermal comfort.

    Bibliography: p.
    1. Environmental engineering (Buildings)
2. Heating.   3. Air conditioning.   I. Title.
TH6021.E4 1975   697   74-13336
ISBN 0-13-166447-6

10   9   8   7

Printed in the United States of America

PRENTICE-HALL INTERNATIONAL, INC., *London*
PRENTICE-HALL OF AUSTRALIA, PTY. LTD., *Sydney*
PRENTICE-HALL OF CANADA, LTD., *Toronto*
PRENTICE-HALL OF INDIA PRIVATE LIMITED, *New Delhi*
PRENTICE-HALL OF JAPAN, INC., *Tokyo*

CONTENTS

FOREWORD

The rapid expansion of environmental technology in the decades since World War II has, ironically, often proved to be as much a liability as an asset for the architect. Theory and equipment alike have proliferated so rapidly that the practitioner has increasingly been compelled to delegate more and more design detailing and specification writing to the specialists. From one point of view, of course, this has been a necessary and useful development. But another and unexpected consequence of this delegation of decision-making to others has been the loss of control by the architect of the overall environmental performance of his buildings. What he needs today is a clearer conceptual grasp of the experiential consequences of environmental manipulation—above all, of the thermal, since it is basic to human comfort.

Professor M. David Egan's book on thermal comfort is a coherent and well-conceived response to this new need. It synthesizes a vast and growing body of data, from a number of separate and sometimes disparate sources, into a form specially designed for the architect. A mere glance at the table of contents will be sufficient to show how comprehensive is his overview of the subject matter and how logically he has been able to organize it. The fact that the material is presented in graphical form (rather than the conventional textual or tabular) is in itself significant, since this coincides with the way in which the architect conceptualizes both problems and solutions. Professor Egan's book should prove of permanent value to students, teachers, and practitioners alike.

JAMES MARSTON FITCH
Professor of Architecture
Graduate School of Architecture and Planning
Columbia University in the City of New York

PREFACE

This book presents the basics of thermal comfort in a graphical format. The verbal descriptions are few, emphasis being on graphical displays to illustrate the significance of climate, materials, and mechanical systems in the design of buildings. It is important, therefore, that the reader carefully peruse all notes on the concept sketches. The sketches are not supplements to the text, but in a very real sense are the text. With reasonable understanding of these concepts, designers (and users) should be better equipped to make mechanical system decisions that will provide for thermal comfort in buildings, the integration of systems with structure and other building services, and the conservation of energy.

The graphical approach also should facilitate understanding of thermal comfort and mechanical system concepts for those in the building professions who have limited time to digest lengthy verbal descriptions. In addition, the tables of technical data should be useful for solving actual building problems. It should be noted, however, that this is *not* a book on how to design the various mechanical systems. The *ASHRAE Handbook & Product Directory* series and some of the selected references given within the text are more appropriate for this engineering design function.

Work on the book began during the author's seminar course in architectural technology at Tulane University. The support for this project by Professor William K. Turner, Dean of the School of Architecture, is appreciated. Mr. Jack H. Brady, Chenault & Brady, Dallas, Texas; Professor Ralph W. Crump, College of Architecture, Cornell University; Professor Richard O. Powell, School of Architecture, Tulane University; and Mr. Joseph J. Quartana, Consulting Engineer, Metairie, Louisiana carefully reviewed the manuscript. Their criticisms and helpful suggestions are gratefully acknowledged. Mr. Quartana's counsel and keen interest in the instruction of mechanical systems are deeply appreciated. Professor Robert D. Eflin, Campus Planner, Clemson University; Dr. Erich A. Farber, Director, Solar Energy Laboratory, University of Florida; Mr. William M.C. Lam, Lam Associates, Cambridge, Massachusetts; and Mr. Jorn Ostergaard, Shell Chemical Company, Deer Park, Texas provided helpful manuscript reviews in their areas of special interest.

Thanks also are due to Dean Harlan E. McClure and the students at the College of Architecture, Clemson University and to Professor Joseph N. Smith and the students at the School of Architecture, Georgia Institute of Technology for their valuable criticisms of the Tulane University edition of this book used by the author in teaching building science and building anatomy courses. Mr. Glen S. LeRoy, Mr. Robert F. Flack and Mr. Knox H. Tumlin, former students at Tulane University, deserve special recognition for their work on example problems and concept sketches. Mr. LeRoy prepared the illustrations for this edition with his customary thoroughness and skill.

Several pages in Section 6 first appeared in the author's book, *Concepts in Architectural Acoustics,* and have been adapted for use with permission of the McGraw-Hill Book Company. A previous publication of Concepts In Thermal Comfort was made by Tulane University. The following is an updated, revised edition.

M. DAVID EGAN, P.E.

INTRODUCTION—HUMAN BODY HEAT LOSS

The human body maintains a balance with its environment through *minor* physiological changes (i.e., by increasing or decreasing the flow of blood to the skin). Body heat losses are primarily by convection, evaporation, and radiation as shown on the sketch below. The basic theory of thermal comfort is presented in Section 1. Note that the thermal comfort factors that influence heat loss are also indicated at the bottom of the sketch.

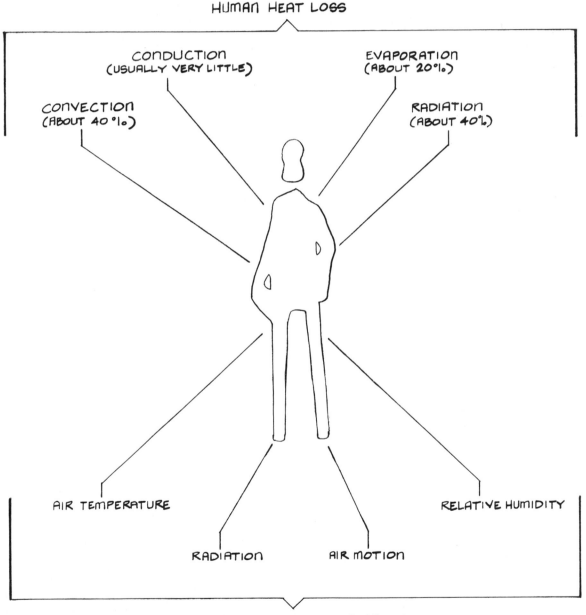

HUMAN HEAT LOSS

CONDUCTION
(USUALLY VERY LITTLE)

EVAPORATION
(ABOUT 20%)

CONVECTION
(ABOUT 40%)

RADIATION
(ABOUT 40%)

AIR TEMPERATURE

RELATIVE HUMIDITY

RADIATION

AIR MOTION

THERMAL COMFORT FACTORS

The designer can achieve thermal comfort conditions in buildings by favorable adaptation of shelter to its climate (Section 2), by proper material selection (Sections 3 and 4), and by effective use of appropriate mechanical systems (Sections 5 and 6).

CONCEPTS IN THERMAL COMFORT

BASIC
THEORY

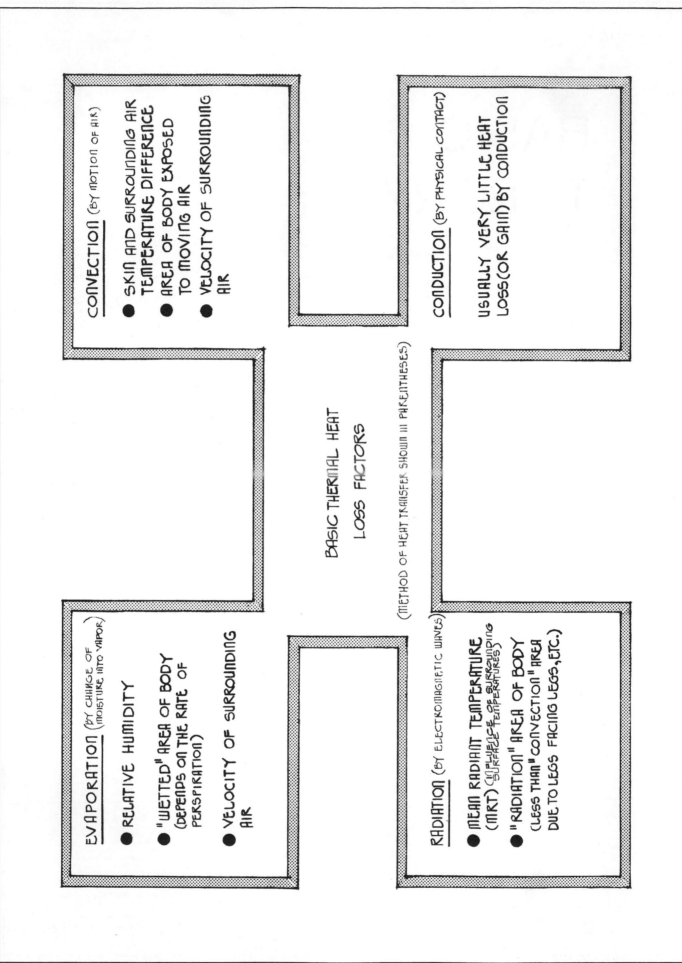

BASIC THERMAL HEAT

LOSS FACTORS

(METHOD OF HEAT TRANSFER SHOWN IN PARENTHESES)

CONVECTION (BY MOTION OF AIR)

● SKIN AND SURROUNDING AIR TEMPERATURE DIFFERENCE
● AREA OF BODY EXPOSED TO MOVING AIR
● VELOCITY OF SURROUNDING AIR

CONDUCTION (BY PHYSICAL CONTACT)

USUALLY VERY LITTLE HEAT LOSS (OR GAIN) BY CONDUCTION

EVAPORATION (BY CHANGE OF MOISTURE INTO VAPOR)

● RELATIVE HUMIDITY
● "WETTED" AREA OF BODY (DEPENDS ON THE RATE OF PERSPIRATION)
● VELOCITY OF SURROUNDING AIR

RADIATION (BY ELECTROMAGNETIC WAVES)

● MEAN RADIANT TEMPERATURE (MRT) (INFLUENCE OF SURROUNDING SURFACE TEMPERATURES)
● "RADIATION" AREA OF BODY (LESS THAN "CONVECTION" AREA DUE TO LEGS FACING LEGS, ETC.)

BASIC THEORY—THERMAL EFFECT OF ACTIVITY

Some common activities are listed below in order of decreasing generated body heat in Btuh. Be sure to include the heat from people in heat gain analyses for summer cooling requirements.

ACTIVITY	BODY HEAT (BTUH)*
① WALKING UP STAIRS	4400
② WALKING DOWN STAIRS	1500
③ WALKING (ABOUT 2 MPH)	750
④ TYPEWRITING	500
⑤ SITTING AT REST (SEE GRAPH ON FOLLOWING PAGE)	400
⑥ SLEEPING	300

* THE SYMBOL "BTUH" MEANS BTU'S PER HOUR. A BRITISH THERMAL UNIT (BTU) IS THE AMOUNT OF HEAT REQUIRED TO RAISE 1 LB. OF WATER 1° FAHRENHEIT (F). NOTE THAT A DEGREE IS A MEASURE OF TEMPERATURE, WHEREAS, A BTU IS A MEASURE OF HEAT ENERGY. FOR EXAMPLE, POUR TWO CUPS OF HOT TEA – ONE FULL AND ONE ⅓ FULL. ALTHOUGH BOTH WILL HAVE THE SAME TEMPERATURE, THE FULL CUP WILL HAVE THREE TIMES AS MUCH HEAT IN BTU'S.

BASIC THEORY—BODY HEAT ADJUSTMENT

Curve "*A*" below shows total body heat losses for seated occupants in a relaxed state. Heat loss by convection and radiation (curve "*B*") and evaporation (curve "*C*") indicate the general way body heat is removed and dissipated at different temperatures. The room relative humidity for the data given below is approximately 45%.

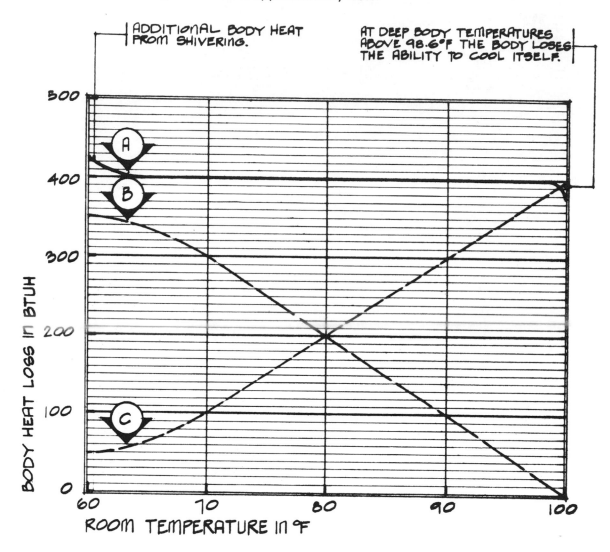

A. *Total body heat generated*

Note that metabolic rate is stable in the temperature range of about 70 to 90°F. Metabolism is the basic physiological process that utilizes energy from food to support vital body functions such as muscle movement, tissue building, etc.

B. *Body heat loss by convection and radiation*

C. *Body heat loss by evaporation*

At low temperatures little heat is dissipated by evaporation as pores tend to close; at high temperatures sweating dissipates heat by evaporative cooling of the skin.

BASIC THEORY—HUMAN BODY TEMPERATURE

Man is a constant temperature animal with a normal deep body temperature of about 98.6°F. Shown below are examples of deep body temperatures caused by extended periods of overheating or overcooling from a wide range of environmental conditions.

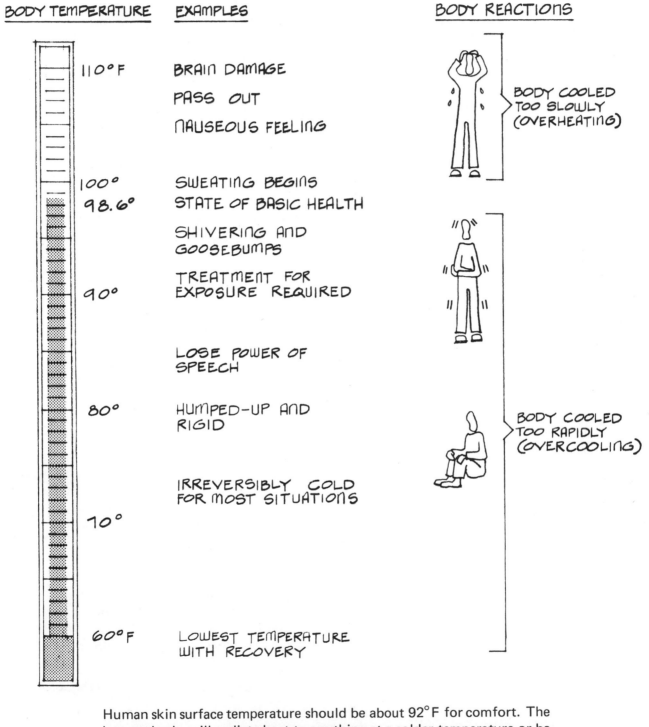

BODY TEMPERATURE EXAMPLES BODY REACTIONS

110°F — BRAIN DAMAGE
PASS OUT
NAUSEOUS FEELING

BODY COOLED TOO SLOWLY (OVERHEATING)

100° — SWEATING BEGINS
98.6° — STATE OF BASIC HEALTH
SHIVERING AND GOOSEBUMPS
90° — TREATMENT FOR EXPOSURE REQUIRED

LOSE POWER OF SPEECH

80° — HUMPED-UP AND RIGID

IRREVERSIBLY COLD FOR MOST SITUATIONS

70°

60°F — LOWEST TEMPERATURE WITH RECOVERY

BODY COOLED TOO RAPIDLY (OVERCOOLING)

Human skin surface temperature should be about 92°F for comfort. The human body will radiate heat to anything at a colder temperature or be heated by anything at a warmer temperature. Building surface temperatures, therefore, are important factors for achieving thermal comfort.

BASIC. THEORY—MEAN RADIANT TEMPERATURE (MRT)

The mean radiant temperature (MRT) is a weighted average of the various radiant influences in a space. It can be estimated by the following formula:

$$\text{MRT} = \frac{\Sigma t\theta}{360} = \frac{t_1\theta_1 + t_2\theta_2 + \ldots t_n\theta_n}{360}$$

where t = surface temperature in °F

           θ = surface exposure angle (relative to occupant) in degrees

Example: An office has exterior concrete block and glass walls having inside surface temperatures of 66°F and 20°F respectively. The interior partitions have 72°F surface temperatures. (Assume that the human body is a cylinder and ignore radiation effects from the floor and ceiling.)

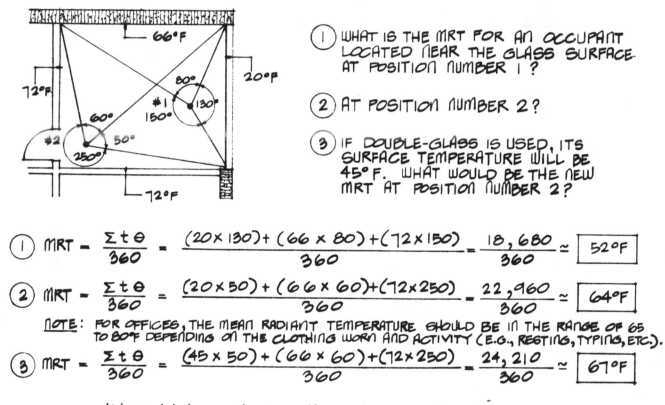

① WHAT IS THE MRT FOR AN OCCUPANT LOCATED NEAR THE GLASS SURFACE AT POSITION NUMBER 1?

② AT POSITION NUMBER 2?

③ IF DOUBLE-GLASS IS USED, ITS SURFACE TEMPERATURE WILL BE 45°F. WHAT WOULD BE THE NEW MRT AT POSITION NUMBER 2?

① $\text{MRT} = \dfrac{\Sigma t\theta}{360} = \dfrac{(20 \times 130) + (66 \times 80) + (72 \times 150)}{360} = \dfrac{18,680}{360} \approx \boxed{52°F}$

② $\text{MRT} = \dfrac{\Sigma t\theta}{360} = \dfrac{(20 \times 50) + (66 \times 60) + (72 \times 250)}{360} = \dfrac{22,960}{360} \approx \boxed{64°F}$

NOTE: FOR OFFICES, THE MEAN RADIANT TEMPERATURE SHOULD BE IN THE RANGE OF 65 TO 80°F DEPENDING ON THE CLOTHING WORN AND ACTIVITY (E.G., RESTING, TYPING, ETC.).

③ $\text{MRT} = \dfrac{\Sigma t\theta}{360} = \dfrac{(45 \times 50) + (66 \times 60) + (72 \times 250)}{360} = \dfrac{24,210}{360} \approx \boxed{67°F}$

It is good design practice to specify exterior constructions that will have inside surface temperatures in winter not more than 5°F below the indoor air temperature. This restriction will result in satisfactory MRT's for most situations. Section 3 presents insulation criteria for controlling heat flow and surface temperatures.

BASIC THEORY—MEAN RADIANT TEMPERATURE CRITERIA FOR ROOMS

The relationship between air temperature in °F and mean radiant temperature (MRT) in °F is shown by the graph below. The shaded region indicates an average zone for human thermal comfort. The graph is based on test data from lightly clothed subjects engaged in sitting activities. Relative humidity is 50% and air velocities range from about 15 to 60 fpm for the data presented.

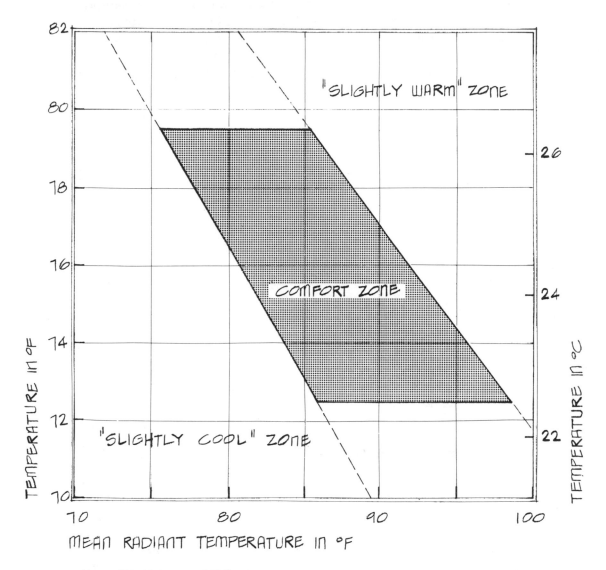

Note: The higher the MRT value, the lower the corresponding air temperature for comfort conditions as indicated by the slope of the comfort zone. Radiant panel heating systems (see Section 5), for example, can provide satisfactory comfort conditions at low air temperatures.

REFERENCE

Fanger, P. O., *Thermal Comfort*. New York: McGraw-Hill, Inc., 1972.

BASIC THEORY—RELATIVE HUMIDITY

Relative humidity (RH) in % is the amount of moisture in air compared to the maximum amount that can exist at a given temperature without condensation. Relative humidity can be measured by a device called a "sling psychrometer." The sling psychrometer shown below consists of two mercury-filled glass thermometers mounted side by side on a frame fitted with a handle by which the device can be whirled through the air. One of the thermometers has a cloth sock that can be wetted. As moisture from the wet sock evaporates, the "wet bulb" temperature lowers. The drier the air surrounding the sling psychrometer, the more moisture that can evaporate from the sock. This evaporation lowers the wet bulb temperature. The greater the difference between the wet bulb and dry bulb temperatures (called "wet bulb depression"), the lower the relative humidity.

Typical Sling Psychrometer

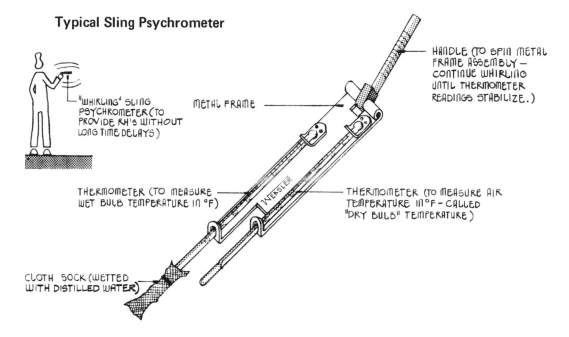

Example Use of Chart to Find RH

Given: Sling psychrometer readings of 75°F dry bulb and 63°F wet bulb. Procedure to find relative humidity from psychrometric chart: Enter dry bulb scale at 75°F and read vertically to closest RH curve at intersection with 63°F diagonal wet bulb scale line (See dashed guidelines below). RH is 50% for air conditions from example sling psychrometer readings.

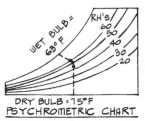

Note: See page 138 for a complete psychrometric chart showing the various properties of air.

BASIC THEORY—TEMPERATURE AND HUMIDITY CRITERIA FOR ROOMS

The relationship between air temperature in °F and relative humidity in % is shown by the graph below. The graph is based on test data from lightly clothed subjects engaged in sitting activities. The shaded region indicates an average zone for human thermal comfort.

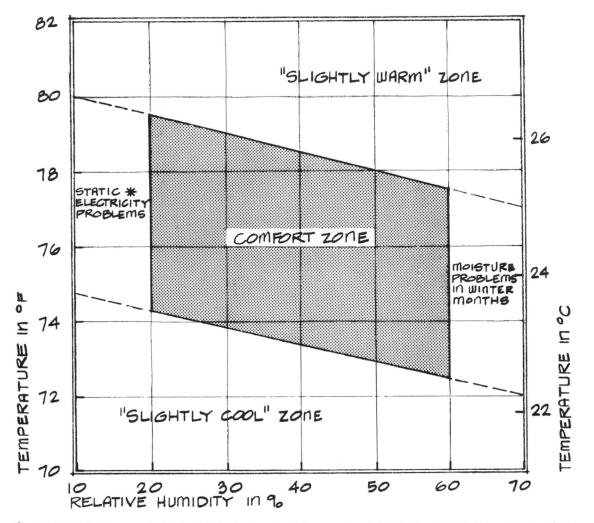

* CARPET IS COMMERCIALLY AVAILABLE THAT HAS A CONDUCTIVE MATERIAL (E.G., COPPER, STAINLESS STEEL, ETC.) WOVEN INTO THE PILES AND BACKING TO HELP REDUCE VOLTAGE BUILDUP.

Note: In the comfort zone, human tolerance to humidity is much greater than tolerance to temperature. Consequently, room air temperatures must be more carefully controlled. Humidity control is also important, however, since high room humidity can cause condensation on glass surfaces in winter, and low room humidity can cause static electricity problems. Typical humidity limitations are indicated in the comfort zone on the above graph. In hospitals careful control of humidity is essential as the level of bacteria propagation is lower in the relative humidity range of 50 to 55%.

REFERENCE

Nevins, R.G., "Criteria for Thermal Comfort." Institute for Environmental Research Report No. 12. Kansas State University.

SUGGESTED AIR TEMPERATURES FOR ROOMS

The suggested temperatures in the table represent guidelines for design based on room heating for winter comfort requirements. For summer conditions, slightly higher temperatures (about +3°F) are appropriate. Human sensitivity to changes in temperature is normally about $\pm 1\frac{1}{2}°$ F. Also note that a temperature of 70°F may be excessively warm for persons exercising in a gymnasium, but the same conditions may be too cool for reading in a residence. Ideally, mechanical systems should have temperature control for each room as occupant comfort requirements can vary with age (older people generally prefer warmer temperatures), sex (very little), clothing worn, and level of physical activity. Requirements can also vary from day to day for the same individual.

Type of Space	*Temperature (in °F)*
Bathrooms, steam and warm-air baths, swimming pools, industrial paint shops, special rooms in hospitals, etc.	75 & above
Residences, hotels, motels, apartments, convalescent homes and homes for the aged, etc.	73 to 75
Courtrooms, churches, classrooms, offices, conference rooms, chapels, hospital patient rooms and wards, etc.	72 to 74
Auditoriums, theaters, large meeting rooms, corridors, lobbies, lounges, cafeterias, restaurants, toilets and service rooms, etc.	68 to 72
Kitchens, laundries, locker rooms, retail shops and stores, hotel ballrooms, etc.	65 to 70
Gymnasiums, exercise rooms, garages, machinery spaces, foundries, factories, industrial shops, etc.	65 & below

BASIC THEORY—TEMPERATURE AND AIR VELOCITY CRITERIA FOR ROOMS

The relationship between moving air stream temperature (i.e., amount above or below room air temperature) in °F and air velocity in feet per minute (fpm) is shown by the graph below. The shaded region indicates an average zone for human thermal comfort.

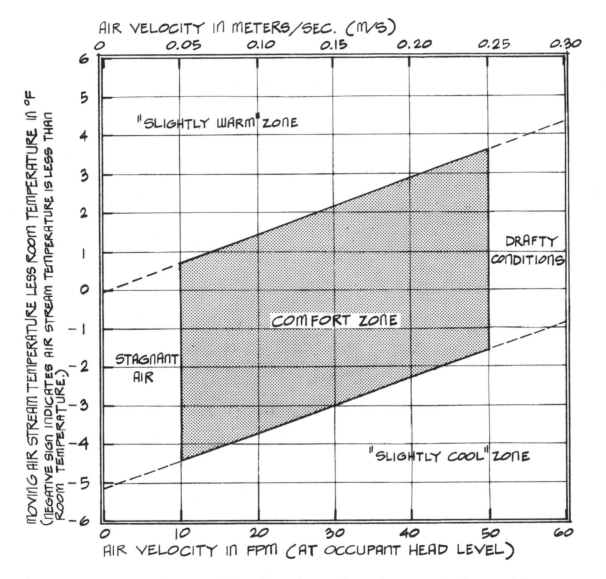

Note: Under some conditions air motion can be a pleasant cooling breeze while under others it can be a draft. Consequently, the combination of air velocity and temperature must be carefully controlled. Typical temperature and velocity limitations are indicated in the comfort zone on the above graph.

REFERENCE

Reinmann, J. J., A. Koestel, and G. L. Tuve, "Evaluation of Three Room Air Distribution Systems for Summer Cooling." *ASHRAE Transactions*, Vol. 65, 1959.

AIR VELOCITY

The table below describes how human occupants react to various conditions of room air motion. Air velocity is expressed in feet per minute (fpm). Moving air contributes to thermal comfort by removing the moisture and heat surrounding the body.

Air Velocity at Head Level of Occupant (in fpm)	Typical Subjective Occupant Evaluation
0 to 10	Complaints about stagnant air.
10 to 50	Generally favorable (manufacturers of air outlet devices, e.g., base performance on 50 fpm air velocity in occupied zone).
50 to 100	Awareness of air motion, but can be comfortable (e.g., some retail shops and stores) when temperature of moving air is above room air temperature.
100 to 200	Constant awareness of air motion, but can be acceptable (e.g., some factories) if air supply is intermittent and above room air temperature.
200 (about 2 mph) and above	Increasingly drafty conditions with complaints about "wind" in disrupting a task, activity, etc.

Note: The combination of air motion and air temperature is an important consideration for achieving comfort. For example, at air velocities of 30 fpm and above, a 15 fpm increase is equal to about a 1°F drop in temperature (See graph on preceding page).

BASIC THEORY—AIR MOTION

Air Stream Motion (Influence of Air Direction)

A guide to desirable air stream directions for seated persons is given below. Note that typical subjective comfort evaluations are indicated for front, rear, side, and overhead air stream directions.

AIR FROM SIDES OR OVERHEAD

AIR FROM FRONT, REAR, OR OVERHEAD

Natural Air Convection (Influence of Glass in Winter)

Air distribution is shown below for two heating air supply locations in a room having a glazed exterior wall. Air supply location can greatly influence occupant comfort and air motion by convection (i.e., tendency of warm air to rise) as shown. The shaded region indicates, in general, how the warm air can blanket room surfaces.

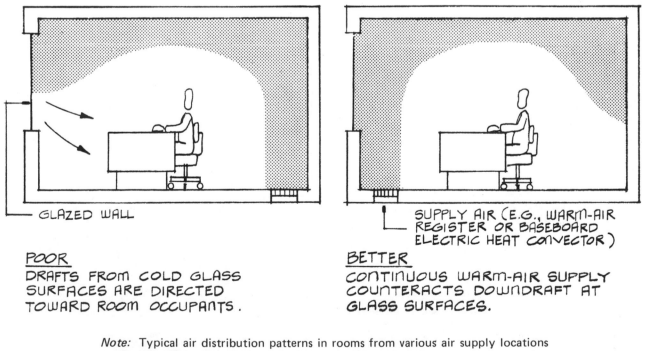

GLAZED WALL

SUPPLY AIR (E.G., WARM-AIR REGISTER OR BASEBOARD ELECTRIC HEAT CONVECTOR)

POOR
DRAFTS FROM COLD GLASS SURFACES ARE DIRECTED TOWARD ROOM OCCUPANTS.

BETTER
CONTINUOUS WARM-AIR SUPPLY COUNTERACTS DOWNDRAFT AT GLASS SURFACES.

Note: Typical air distribution patterns in rooms from various air supply locations for both heating and cooling are shown on the following page.

BASIC THEORY—AIR DISTRIBUTION IN ROOMS

The best supply air outlet locations for heating, as shown by the distribution patterns below, are located near the floor (preferably at outside walls under glass areas to prevent cold downdrafts). For cooling, however, the best supply air outlet locations are high on inside walls or in the ceiling. Consequently, the primary comfort requirement of the climate region—heating or cooling—should influence supply air outlet location. The shaded areas below indicate the distribution pattern for the primary air (i.e., supply air plus entrained air near the outlet location) and room air under the influence of outlet conditions, having velocities in the range of 35 to 150 fpm. The arrows indicate convection currents caused by the difference in temperature between room air and enclosure surface temperatures.

Supply Air Outlet Located High on Inside Wall (or at Ceiling)

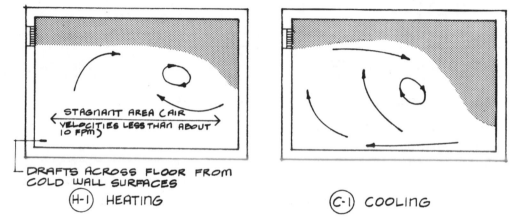

(H-1) HEATING

(C-1) COOLING

Supply Air Outlet Located at Floor (or Sill)

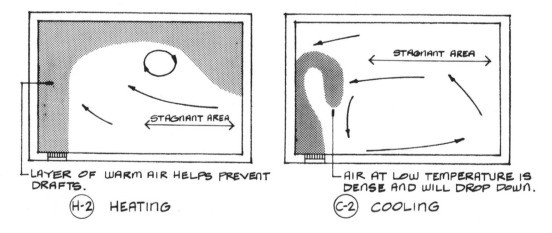

(H-2) HEATING

(C-2) COOLING

Note: Generally, the return air location has little effect on room air distribution. However, try to avoid floor return air locations as they collect dirt too easily and can conflict with room furnishings.

REFERENCE

"Room Air Distribution Considerations." National Warm Air Heating and Air Conditioning Association, 1st Ed., *Manual E* (1965).

SUGGESTED AIR CHANGES FOR ROOMS

Type of Space	Air Change (per hr.)
Machinery spaces, maintenance shops (such as for electrical equipment), school and industrial shops, etc.	8 to 12
Cafeterias, restaurants, offices, reception areas, hospitals, residences, garages, etc.	6 to 20
Churches, libraries, reading rooms, bowling alleys, retail shops and stores, etc.	15 to 25
Auditoriums, theaters, classrooms, kitchens, conference rooms, etc.	10 to 30

The table above suggests ranges of room air change (i.e., complete displacement of room air) for acceptable air motion. The appropriate air change value depends on the number and activity of room occupants. Note that the values given in the table are also well above the requirements to replace the oxygen consumed in breathing (2 cfm per person) and to control the carbon dioxide concentrations in the air (3 cfm per person). Air change is related to airflow volume in cubic feet per minute (cfm) by the formula:

$$N = 60 \frac{Q}{V}$$

where N = number of air changes per hour (ach)

Q = airflow volume in cfm

V = room volume in cubic feet

Room air change can contribute to comfort and condensation control (See Section 3) by removing heat and moisture produced by room activities.

Example: A small conference room for an insurance company is 24 ft by 12 ft in plan with 9 ft ceiling height. Assume total room supply airflow volume (Q) of 500 cfm. What would be the number of air changes (N) per hour? First, find the volume (V) of the room:

$$V = 24 \times 12 \times 9 = 2592 \text{ cu ft}$$

next,
$$N = 60 \frac{Q}{V} = 60 \frac{500}{2592} \simeq 12 \text{ changes/hr}$$

$$(\simeq \text{means approximately equal to})$$

Conference rooms should be in the range of 10 to 30 changes per hour so air change requirements are satisfied.

Note: Airflow volumes for odor control may be achieved by outdoor air and/or by air that has been cleared of odors by spray washers, space-charge neutralization, or other means. See page 119 for clean, odor-free outdoor air requirements for various applications presented in terms of occupant smoking activity.

BASIC THEORY—SOURCES OF MOISTURE IN BUILDINGS

Moisture control in buildings is important as high humidity can cause condensation on glass surfaces in winter. Also, moisture from building occupants, equipment, etc. influences the size of refrigeration elements required for summer cooling because moisture will change the condition of the air. Be careful to provide exhaust systems to directly ventilate spaces such as kitchens, exercise rooms, toilet rooms, etc. for odor control as well as moisture control and heat removal.

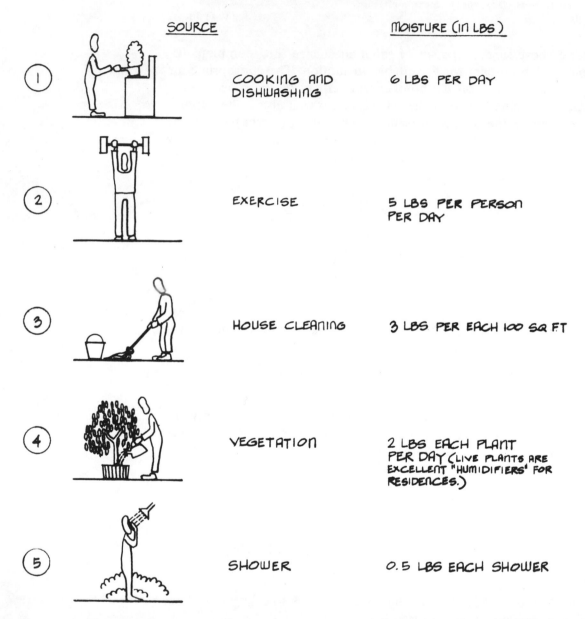

	SOURCE	MOISTURE (IN LBS)
1	COOKING AND DISHWASHING	6 LBS PER DAY
2	EXERCISE	5 LBS PER PERSON PER DAY
3	HOUSE CLEANING	3 LBS PER EACH 100 SQ. FT
4	VEGETATION	2 LBS EACH PLANT PER DAY (LIVE PLANTS ARE EXCELLENT "HUMIDIFIERS" FOR RESIDENCES.)
5	SHOWER	0.5 LBS EACH SHOWER

High humidity retards human heat loss by evaporative cooling (i.e., sweating) and by respiration. Low humidity tends to dry throat and nasal passages. In buildings, low humidity can also cause loosened furniture joints, cracked book bindings, etc. In hospital operating rooms do *not* allow low humidity as a spark could explode anesthetic gases. Humidification devices are available that can introduce moisture directly into the air stream of mechanical systems.

MOISTURE CONTENT OF AIR

The table below gives moisture content of air at standard conditions in pounds per *1000* cubic feet. It shows the amount of moisture needed to maintain a particular relative humidity at various dry bulb temperatures.

Temperature (°F)	Relative Humidity (%)									
	20	25	30	35	40	45	50	55	60	70
0	0.012	0.015	0.018	0.021	0.024	0.027	0.029	0.032	0.035	0.041
10	0.020	0.025	0.029	0.034	0.039	0.044	0.049	0.054	0.059	0.069
20	0.031	0.039	0.047	0.055	0.063	0.071	0.079	0.087	0.095	0.110
30	0.051	0.064	0.077	0.089	0.103	0.116	0.128	0.141	0.154	0.180
40	0.077	0.097	0.117	0.135	0.155	0.174	0.193	0.213	0.233	0.272
50	0.123	0.141	0.171	0.198	0.227	0.259	0.284	0.312	0.341	0.399
60	0.162	0.205	0.246	0.288	0.329	0.370	0.412	0.453	0.495	0.579
70	0.232	0.291	0.349	0.408	0.467	0.527	0.585	0.645	0.704	0.824
72	0.249	0.312	0.375	0.438	0.501	0.564	0.628	0.691	0.755	0.884
74	0.267	0.334	0.401	0.469	0.536	0.605	0.671	0.740	0.809	0.947
76	0.285	0.357	0.429	0.502	0.574	0.647	0.720	0.792	0.866	1.012
78	0.305	0.382	0.458	0.535	0.614	0.692	0.769	0.848	0.925	1.082
80	0.326	0.408	0.490	0.574	0.656	0.740	0.823	0.906	0.991	1.160
86	0.389	0.496	0.597	0.698	0.799	0.901	1.003	1.105	1.212	1.418
90	0.450	0.563	0.678	0.793	0.908	1.024	1.141	1.258	1.375	1.613

The table can also be used to determine how much moisture should be added to, or removed from, the air supply system to achieve design relative humidity. This is found by subtracting the outdoor air moisture content from the moisture content at indoor design conditions.

Example: A data processing equipment room requires a relative humidity maintained uniformly at approximately 50% to help preserve paper quality, dimensional stability for process cards and tapes, etc. If outdoor air design conditions are R.H. of 25% and dry bulb temperature of 30°F, how much moisture must be added to this air to achieve 50% R.H.? Assume indoor design air temperature of 70° F.

$$0.585 \text{ lbs/1000 cu ft at } 70°F \ \& \ 50\% \text{ R.H.}$$
$$\underline{-0.064} \text{ lbs/1000 cu ft at } 30°F \ \& \ 25\% \text{ R.H.}$$

Required moisture = 0.521 lbs/1000 cu ft

The human comfort zone for relative humidity is from 20 to 60%. Nevertheless, some industrial applications (e.g., textile mills, optical lens grinding, food storage) require R.H.'s greater than 60% due to the equipment used, manufacturing process, or kinds of products stored. At the other extreme, some applications (e.g., certain pharmaceutical products, plywood cold pressing) require R.H.'s less than 20%.

CLIMATE AND SHELTER CONSIDERATIONS

CLIMATE AND SHELTER CONSIDERATIONS—CLIMATE CHART

Effects of climate for moderate climate regions in the United States are shown below. The comfort zone indicates conditions where persons will feel comfortable in the shade. When conditions of temperature and relative humidity fall outside the comfort zone, corrective measures— winds, sunshine, or moisture—can produce comfortable conditions. Prevailing climate conditions for a specific location can be plotted on the chart to find what corrective measures are needed. Many may be achieved by natural means (e.g., landscaping, structure openings, and orientation), whereas, others require mechanical air-conditioning.

REFERENCE

Olgyay, V., *Design With Climate*. Princeton, New Jersey: Princeton University Press, 1963, p. 22.

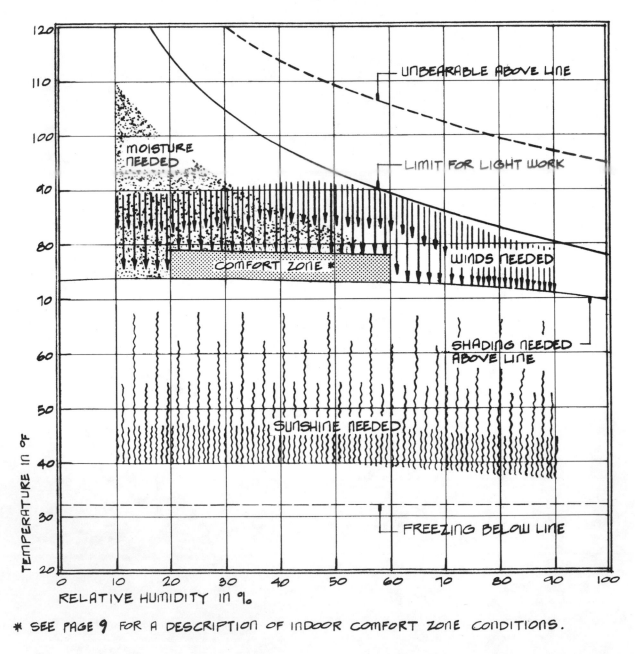

* SEE PAGE 9 FOR A DESCRIPTION OF INDOOR COMFORT ZONE CONDITIONS.

CLIMATE AND SHELTER CONSIDERATIONS—CLIMATE REGIONS

The four major climate regions—cool, temperate, hot-arid, and hot-humid—in the United States are shown below. Also, the north latitudes indicated on the map can be used to help determine the appropriate solar heat gain table for many locations.

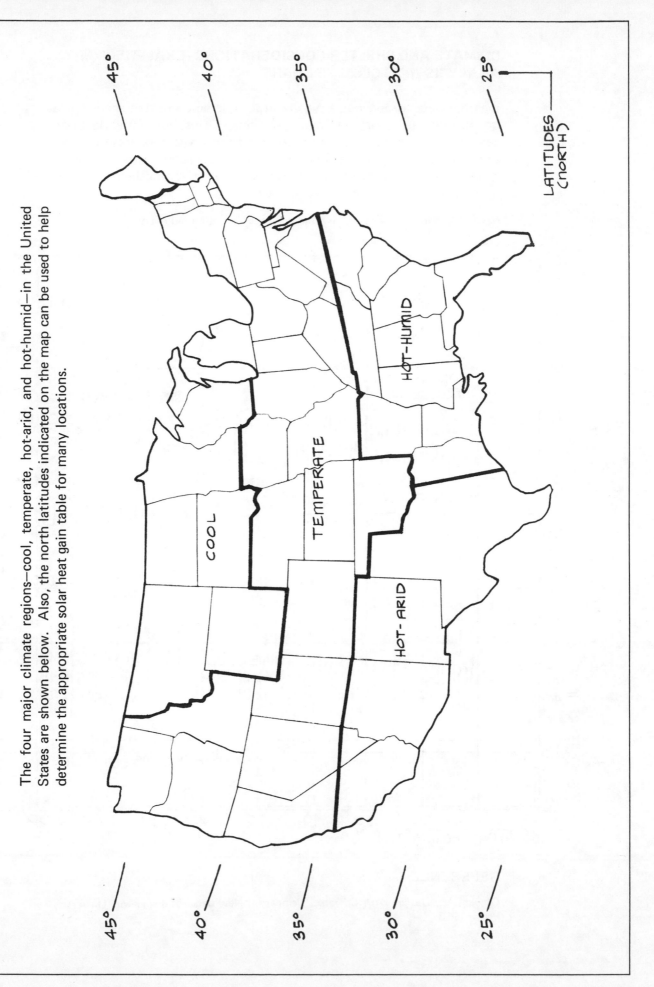

CLIMATE AND SHELTER CONSIDERATIONS—EXAMPLE ANALYSIS WITH CLIMATE CHART

Weather data throughout a typical year is depicted by the shaded area on the climate chart for Miami, Florida. This hot-humid location generally has high relative humidity conditions with moderate temperature variations. The chart indicates that shading and winds are needed most of the year. Suggested building design objectives for the hot-humid region are shown on the following page.

Note: See page 25 for optimum building shapes in hot-humid region.

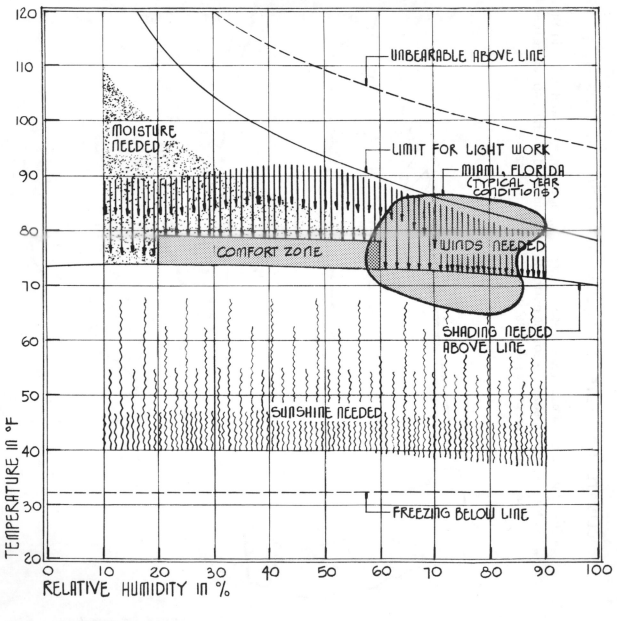

REFERENCE

Olgyay, V., *Design With Climate*. Princeton, New Jersey: Princeton University Press, 1963.

CLIMATE AND SHELTER CONSIDERATIONS—EXAMPLE STRUCTURE FOR HOT-HUMID REGION

The example structure for the hot-humid region is elongated in the east-west direction. Design objectives are to: 1. reduce penetration of solar radiation, 2. remove inside heat from people, lights, etc., and 3. improve evaporative cooling conditions by natural ventilation. Air motion through window openings (oriented to intercept prevailing summer breezes) and roof ridge vent is shown by arrows on the sketch below. For effective attic ventilation, air should also enter through low inlets located in the soffits of the roof overhang.

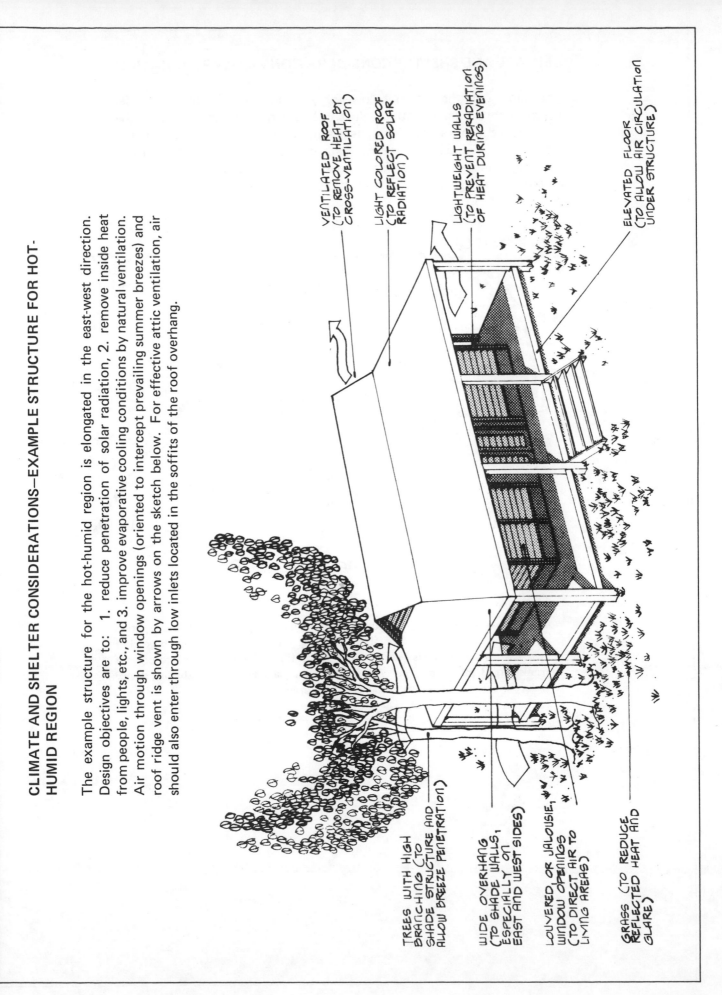

VENTILATED ROOF (TO REMOVE HEAT BY CROSS-VENTILATION)

LIGHT COLORED ROOF (TO REFLECT SOLAR RADIATION)

LIGHTWEIGHT WALLS (TO PREVENT RERADIATION OF HEAT DURING EVENINGS)

ELEVATED FLOOR (TO ALLOW AIR CIRCULATION UNDER STRUCTURE)

TREES WITH HIGH BRANCHING (TO SHADE STRUCTURE AND ALLOW BREEZE PENETRATION)

WIDE OVERHANG (TO SHADE WALLS, ESPECIALLY ON EAST AND WEST SIDES)

LOUVERED OR JALOUSIE, WINDOW OPENINGS (TO DIRECT AIR TO LIVING AREAS)

GRASS (TO REDUCE REFLECTED HEAT AND GLARE)

CLIMATE AND SHELTER CONSIDERATIONS—SOLAR RADIATION

The sun's orbit is shown below for 21 June at 40° north latitude. Approximate solar radiation values in Btuh/sq ft are indicated on the exaggerated orbit sketch for a clear atmosphere at morning, noon, and afternoon. Typical solar design data is given on page 34 for various building orientations at different seasons and times of day. The orbit of the sun across the sky at a location will vary in angle and elevation with the season.

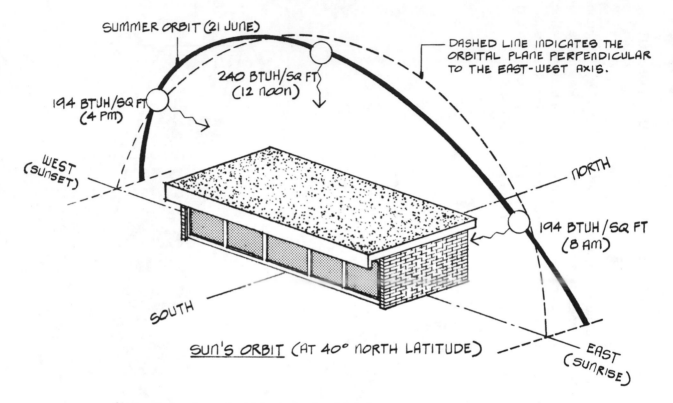

SUMMER ORBIT (21 JUNE)

DASHED LINE INDICATES THE ORBITAL PLANE PERPENDICULAR TO THE EAST-WEST AXIS.

240 BTUH/SQ FT (12 noon)

194 BTUH/SQ FT (4 PM)

194 BTUH/SQ FT (8 AM)

WEST (SUNSET)

NORTH

SOUTH

EAST (SUNRISE)

SUN'S ORBIT (AT 40° NORTH LATITUDE)

Note: If glass is used within the sun's orbit it should be effectively shaded by trees or external shading devices. The above building, for example, does not have glass on the east and west walls because the construction is concrete block to prevent transmission of solar radiation. The south wall is glass with a panel shading device to reduce the transmitted solar radiation at that orientation.

CLIMATE AND SHELTER CONSIDERATIONS—BUILDING SHAPES

Studies of square and various oblong shapes and orientations of buildings in the four major climate regions in the United States show that the shapes below are preferred. The objective of this study by Olgyay was to find a balance between the underheated season, when solar radiation can be beneficial, and the overheated season, when radiation should be avoided. Note that the optimum shape in all regions is elongated in the east-west direction.

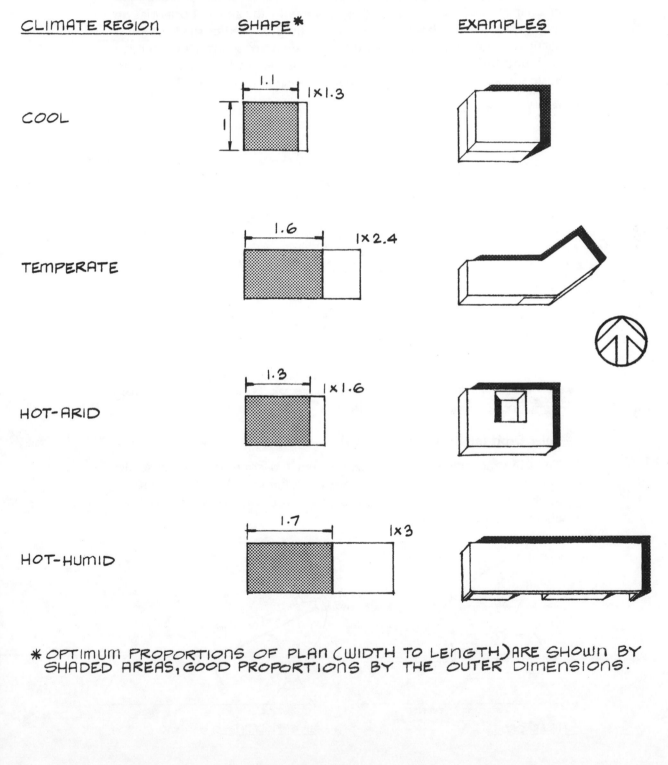

CLIMATE REGION — SHAPE* — EXAMPLES

COOL — 1.1 / 1 — 1×1.3

TEMPERATE — 1.6 — 1×2.4

HOT-ARID — 1.3 — 1×1.6

HOT-HUMID — 1.7 — 1×3

* OPTIMUM PROPORTIONS OF PLAN (WIDTH TO LENGTH) ARE SHOWN BY SHADED AREAS, GOOD PROPORTIONS BY THE OUTER DIMENSIONS.

CLIMATE AND SHELTER CONSIDERATIONS—SOLAR GEOMETRY

The position of the sun in relation to specific geographic location, season, and time of day can be found by geometric techniques. Shading effects from external building elements, nearby structures, etc. can also be studied with scale models or heliodon devices.

Sun Angles

Bearing (or azimuth) angle (β) and altitude angle (α) of the sun are given in solar tables.* These sun angles, along with the angle of orientation relative to the north-south axis, can be used to predict shadows for a particular time at a specific latitude. The depth of shade (d) in feet can be found by $d = x$ ($\tan \alpha/\cos \beta$) where x is the overhang width in feet. Basic sun angle geometry is given below.

*See: pp. 388-92. ASHRAE Handbook of Fundamentals (1972).

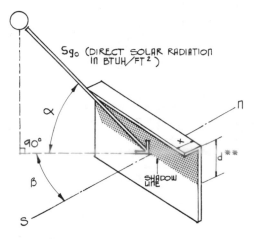

VERTICAL SURFACE (WITH OVERHANG)

** SEE PAGE 31 FOR A TABLE OF APPROXIMATE d VALUES FOR VARIOUS GLASS EXPOSURES IN TERMS OF NORTH LATITUDE IN DEGREES. ALSO SEE SHADING CHART ON P. 57, CARRIER SYSTEM DESIGN MANUAL, PART #1 (1960).

Solar Orbit (at 40° North Latitude)

The sun's movement across the sky differs in bearing and altitude angles with the seasons. Exaggerated orbits for summer and winter conditions at 40° north latitude are shown below. Note that seasonal variations also alter the solar radiation values.

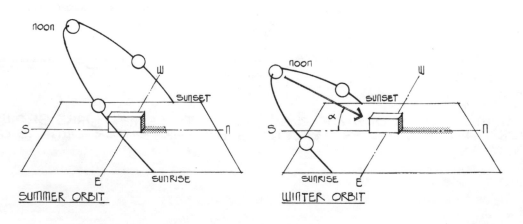

SUMMER ORBIT   WINTER ORBIT

CLIMATE AND SHELTER CONSIDERATIONS—SHADING FROM BUILDINGS

Buildings can provide useful shade on nearby structures. For example, the planned structure, located at 40° north latitude, will be shaded as shown below on the afternoon of 23 July. However, do *not* rely on shade from buildings near a proposed site unless it is certain that they will be preserved in their existing conditions.

Elevation View of Buildings (Showing Shading at 40° North Latitude)

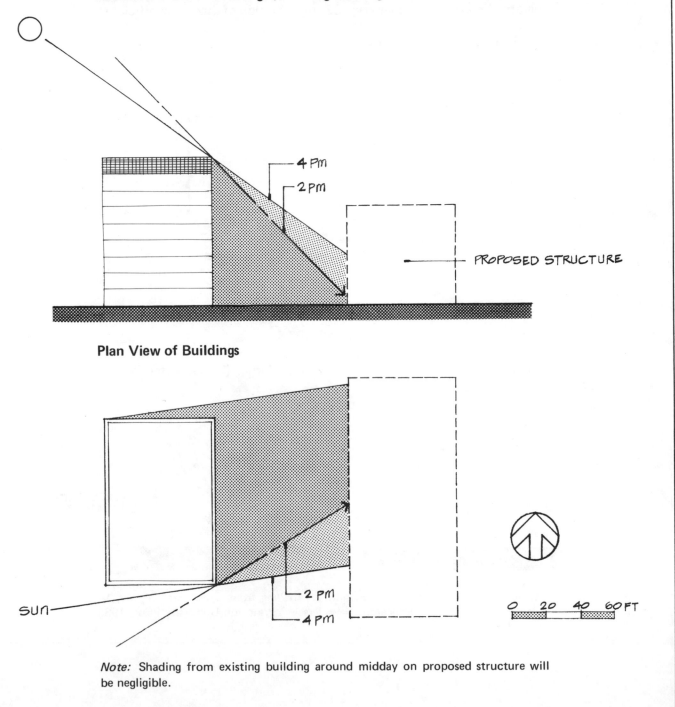

4 PM

2 PM

PROPOSED STRUCTURE

Plan View of Buildings

sun

2 PM

4 PM

0 20 40 60 FT

Note: Shading from existing building around midday on proposed structure will be negligible.

CLIMATE AND SHELTER CONSIDERATIONS—SHADING FROM TREES AND VEGETATION

Trees and vegetation can be used to provide shade where it will be seasonally beneficial. For example, deciduous trees (i.e., ash, elm, maple, poplar) are especially effective because they provide shade during the overheated periods of the summer months and, by shedding their leaves in winter, do not interfere with useful solar radiation. When properly placed, mature trees have shading coefficients (*S.C.*) from 0.25 to 0.20. Shown below is an effective shading layout for a building in the temperate climate region. (See page 32 for definition of shading coefficient.)

Typical Building Shading Layout

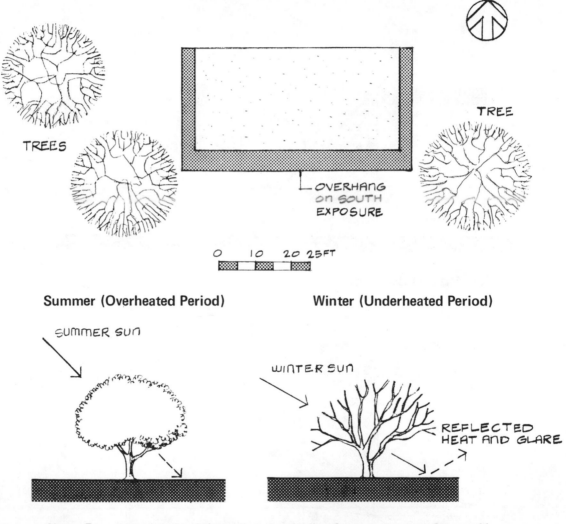

Summer (Overheated Period)   Winter (Underheated Period)

Note: For comprehensive information on shading from trees, see: Olgyay, V., *Design With Climate*. Princeton, New Jersey: Princeton University Press, 1963, pp. 74-7.

CLIMATE AND SHELTER CONSIDERATIONS—SHADING DEVICES

The horizontal exterior solar shading devices shown below are primarily useful for south exposures. East and west exposures will require excessive width of overhang to be effective. (See page on shade from overhangs for design guidelines.)

Cantilever (or Overhang)

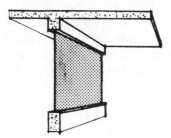

GOOD

Louver Overhang (Horizontal)

Note: The open construction between louvers prevents hot air buildup near walls.

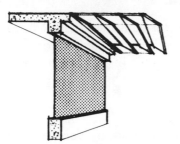

Panels (or Awnings)

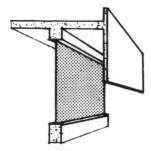

Horizontal Louver Screen

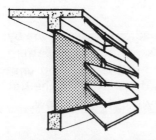

BEST

Note: Check glass manufacturer's installation recommendations carefully when heat-absorbing glass is used as exterior shading devices, balconies, etc. can cause undesirable tension stresses on glass edges.

CLIMATE AND SHELTER CONSIDERATIONS—SHADING DEVICES

The vertical exterior louver and egg-crate solar shading devices shown below are primarily useful for east and west exposures. These devices will also improve the insulation value of glass in winter months by acting as a windbreak.

Vertical Louvers

Vertical elements can also be designed to vary their angle according to the sun's position. Moveable vertical louvers can provide *S.C.*'s from 0.15 to 0.10. Due to problems from icing, they are not practical in cold regions. On cloudy days, a photocell control device can set moveable louvers to the perpendicular position shown below for maximum light penetration.

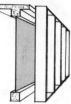

Indoor louvers with integral tubing can be water-cooled to help remove heat from solar radiation at glass areas. In the underheated season, the louvers also can be heated by a warm water circuit (e.g., heat removed from lighting fixtures) to prevent cold drafts.

Egg-Crate

The egg-crate solar shading device is a combination of vertical and horizontal elements. Because of its high shading efficiency (e.g., *S.C.* ≤0.10), egg-crate devices are often used in hot climate regions. Note also that the horizontal elements control ground glare from reflected solar rays.

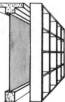

Note: For additional information on shading devices, see: Olgyay, V., *Design With Climate.* Princeton, New Jersey: Princeton University Press, 1963, p. 68-71.

Walls

The east and west sides of a building can be shaped to provide shade by facing glass away from the sun. For example, a "sawtooth" wall pattern (typical shape for industrial roof skylights) used in the cool and temperate regions should concentrate glass towards the south. In the hot regions, glazing should face the north as shown by the sketch below.

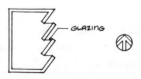

GLAZING

CLIMATE AND SHELTER CONSIDERATIONS—SHADE FROM OVERHANGS

The table below can be used to determine the approximate amount of shade from overhangs. The values in the table are averages for the 5 hours of maximum solar intensity on 1 August for a given wall orientation. To find the depth of shade (d), multiply the width of overhang (x) by the value (k) from the table. For example, at 40° north latitude an overhang on the southeast exposure having a width of 8 ft will provide a depth of shade about 1.25 X 8 = 10 ft. (Enter table at 40° north latitude row and read k = 1.25 at intersection of southeast exposure column.)

$$d = kx$$

where x = width of overhang in feet
         d = depth of shade in feet
         k = multiplier value from table (no units)

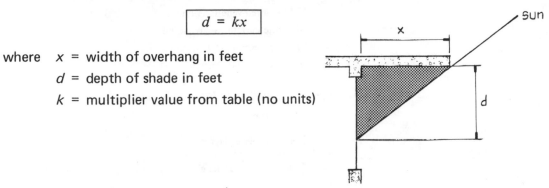

k Value Table (For Formula Above)

NORTH LATITUDE IN DEGREES	GLASS EXPOSURE			
	E & W	SE & SW	NW, N & NE	S
50	0.79	1.01	1.24	1.70
45	0.80	1.13	1.44	2.05
40	0.81	1.25	1.67	2.60
35	0.82	1.41	1.79	3.55
30	0.83	1.63	2.89	5.40
25	0.83	1.89	4.63	10.10

Note: It is good practice to provide a depth of shade covering the total glass area under the overhang.

SHADING COEFFICIENTS

The effectiveness of any shading device can be expressed by its shading coefficient (*S.C.*). This coefficient describes the fraction of the incident solar energy that is transmitted. Theoretically, it can vary from 1.0 (all solar energy transmitted, e.g., unshaded, clear window glass) to 0 (no solar energy transmitted).

Description	S.C.
Interior:	
1. Aluminum venetian blinds (with slats adjusted to prevent transmission of direct sun rays)	0.45
2. Light-colored venetian blinds	0.55
3. Off-white curtain (6-8 oz/sq yd)	0.40
4. Dark gray curtain (6-8 oz/sq yd)	0.60
5. Open-weave fabric curtain	0.75 to 0.60
Glass (With no Shading Indoors or Outdoors):	
6. $\frac{1}{4}$ in. clear plate-glass	0.95
7. $\frac{3}{8}$ in. clear plate-glass	0.90
8. $\frac{3}{16}$ in. heat-absorbing glass (gray, bronze, or green tinted)	0.75
9. $\frac{1}{2}$ in. heat-absorbing glass	0.50
10. Dark gray metallized reflective coating on glass	0.35 to 0.20
11. Light gray metallized reflective coating on glass	0.60 to 0.35

Note: See manufacturer's literature for precise data and characteristics of a given product. For example, warm air from floor registers directed at heat-absorbing glass in winter can cause breakage from undesirable tension stresses on glass edges. Locate registers on room side of interior shading devices (e.g., draperies). Be careful also to avoid using reflective glass on exposures that could reflect solar energy towards nearby glazed buildings.

Landscaping:	
12. Mature tree that provides dense shading	0.25 to 0.20
13. Young tree that provides light shading	0.60 to 0.50

Note: Use caution when estimating shade from trees. For example, trees can die from disease and deciduous trees will provide almost no shading during the months when their leaves have fallen.

External Elements (See pages 29 & 30.):	
14. Egg-crate	0.10
15. Panels or awnings (light color)	0.15
16. Horizontal louver overhang	0.20
17. Horizontal louver screen	0.60 to 0.10
18. Continuous overhang or cantilever	0.25
19. Vertical louvers or fins (fixed)	0.30
20. Vertical louvers or fins (moveable)	0.15 to 0.10

Skylights:

21.	Clear acrylic plastic dome	0.95 to 0.50
22.	Gray acrylic plastic dome	0.90 to 0.45
23.	Bronze acrylic plastic dome	0.90 to 0.35

LOW-RISE DOME

When more than one type of shading device occurs on an exposure, use the lowest *S.C.* value for the devices provided. For example, if a horizontal louver overhang, having an *S.C.* of 0.20, is used on an exposure that also has several young trees nearby with an *S.C.* of about 0.55, use the 0.20 value for design computations. For more information on shading devices, see: Ramsey, C.G., and H.R. Sleeper, *Architectural Graphic Standards.* N.Y.: John Wiley & Sons, Inc., 1971.

TYPICAL SOLAR DESIGN DATA

Approximate solar radiation (S_g) is given below at the 40° and 32° north latitudes. Typical cloudless day conditions in Btuh/ft² are listed for 21 December, 21 March, 21 September, and 21 June. In heat gain analyses select the time of day at which the solar heat gain will be greatest. For example, a building with glass only on the west exposure will have a peak heat gain at 4 p.m. In most situations, however, it will be necessary to do additional computations to find the maximum solar heat gain.

APPROXIMATE SOLAR RADIATION THROUGH CLEAR, UNSHADED GLASS (Btuh/ft²)

Exposure 40° North Latitude:	21 December 8 a.m.	Noon	4 p.m.	21 March & 21 Sept. 8 a.m.	Noon	4 p.m.	21 June 8 a.m.	Noon	4 p.m.
N	2	15	2	14	27	14	26	34	26
NE	6	15	2	82	27	14	140	34	23
E	60	16	2	196	29	14	194	37	23
SE	75	159	3	190	130	14	137	64	23
S	44	228	44	66	185	66	26	85	26
SW	3	159	75	14	130	190	23	64	137
W	2	16	60	14	29	196	23	37	194
NW	2	15	6	14	27	82	23	34	140
Roof	6	102	6	76	201	76	138	240	138
Alt. angle (α)	6	27	6	23	50	23	37	74	37
Bearing angle (β)*	53	0	53	70	0	70	91	0	91
32° North Latitude:									
N	7	19	7	16	29	16	32	36	32
NE	16	19	7	96	29	16	154	37	23
E	121	21	7	204	31	16	193	38	23
SE	149	159	7	188	110	16	121	47	23
S	86	227	86	56	158	56	25	54	25
SW	7	159	149	16	110	188	23	47	121
W	7	21	121	16	31	204	23	38	193
NW	7	19	16	16	29	96	23	37	154
Roof	20	142	20	89	227	89	136	248	136
Alt. Angle (α)	10	35	10	25	58	25	37	82	37
Bearing angle (β)*	54	0	54	73	0	73	97	0	97

*Bearing angle is angle in horizontal plane measured from the south.

The radiation values above do not represent the maximum values that can occur, but are typical for average cloudless day conditions with moderate levels of smoke and dust in the atmosphere. For detailed solar design data at various latitudes, seasons, and times of day, see: *ASHRAE Handbook of Fundamentals* (1967 or 1972 editions) or *Trane Solar Table Manual* (1966).

USE OF SOLAR DESIGN DATA

Example: Find the solar heat gain through the glazed openings of a building located near 40° north latitude. Evaluate summer conditions on 21 June at 8 a.m., noon, and 4 p.m. There is a 40 ft by 10 ft high glass curtain wall on the south exposure with a continuous 4 ft wide overhang. In addition, 100 sq ft of glass on the west exposure is shaded by 8 oz/sq yd dark gray curtains. Shading coefficients (*S.C.*'s), defined on page 32, are as follows:

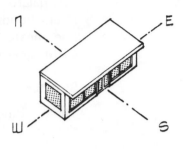

$$S.C. = 0.25 \text{ (for continuous overhang on S exposure)}$$
$$S.C. = 0.60 \text{ (for curtains on W exposure)}$$

The time of day at which the solar heat gain will be greatest is computed below.

Time	Exposure	S_g	A(area in ft²)	Heat Gain (H.G.) without shading*	S.C.	H_o*
8 a.m.	S	26	400	10,400	0.25	2,600
	W	23	100	2,300	0.60	1,380
				12,700		3,980
12 noon	S	85	400	34,000	0.25	8,500
	W	37	100	3,700	0.60	2,220
				37,700**		10,720
4 p.m.	S	26	400	10,400	0.25	2,600
	W	194	100	19,400	0.60	11,640
				29,800		14,240**

*H.G. = A S_g and H_o = A S_g (S.C.) in Btuh. (See page 78 for the basic heat gain formulas.)

**Note that the maximum solar heat gain through unshaded glass occurs at noon; whereas, the maximum heat gain using an overhang on the south and curtains on the west occurs at 4 p.m. Also, the shading provided reduces the solar heat gain by about 60% for this example problem.

CLIMATE AND SHELTER CONSIDERATIONS—MACRO-SHADING DEVICES

The macro-(or large scale) shading devices shown below can be used to shield buildings from solar radiation in the overheated season. In the underheated season, the building can be exposed to the sun to provide useful solar radiation.

Structures

The building structure can be rotated in response to the movement of the sun.

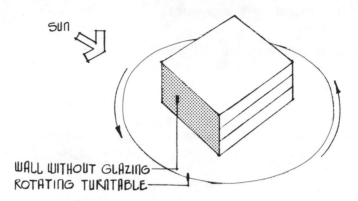

Barriers

Barrier screens can be moved about buildings in response to the movement of the sun. This kind of shading device also can serve as a windbreak.

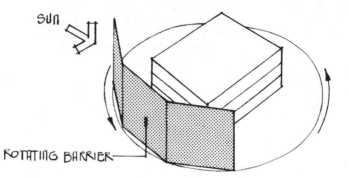

Note: The cost of turning devices for the above examples is the limiting economic consideration at the present time.

REFERENCE

Fitch, J. M., "The Control of the Luminous Environment." *Scientific American,* Vol. 219, No. 3, September 1968.

CLIMATE AND SHELTER CONSIDERATIONS—SUN ORIENTATIONS FOR ROOMS

Desirable sun orientations for various rooms in residences (located above 35° north latitude) can be found in the table below. The circles indicate orientations, at the top of the columns, suitable for a given activity. Solar radiation can contribute to thermal comfort in winter and can be attenuated in summer by shading devices (e.g., trees, overhangs, etc.). In addition, the psychological desire of occupants for contact with the outdoors and the germicidal action of radiation are important considerations.

	N	NE	E	SE	S	SW	W	NW
BEDROOM	●	●	●	●	●	●		
LIVING ROOM				●	●	●	●	
KITCHEN			●	●	●	●		
BATHROOM	●	●	●	●	●	●	●	●
FAMILY ROOM				●	●	●	●	
LIBRARY	●	●						●
WORKSHOP	●	●						●

REFERENCE

Aronin, J., *Climate and Architecture*. New York: Reinhold Publishing Corporation, 1953.

CLIMATE AND SHELTER CONSIDERATIONS—NATURAL VENTILATION

Shown below are airflow patterns from wind through rooms having various inlet locations and exterior shading devices. In general, the location of the outlets does not affect airflow pattern. To achieve beneficial air motion in summer months, inlet openings should be located to intercept prevailing summer breezes, outlets should be equal or larger than inlets in area, and interior open plans should be used. Air motion should be directed toward the occupied (or living) area. Note that the use of insect screens will reduce the airflow into buildings.

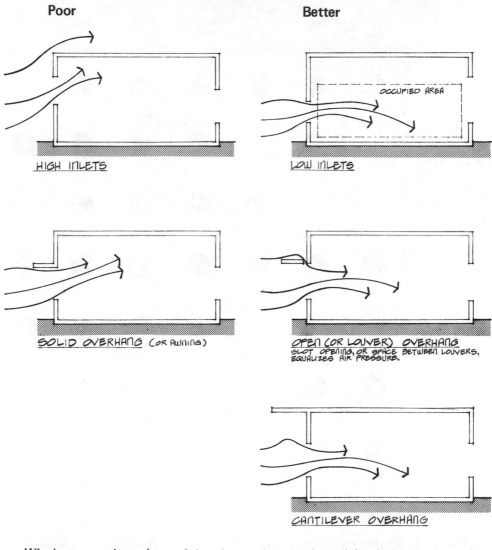

Window openings (e.g., jalousie, or louvered, and horizontal pivoted types) can be adjusted to help direct air to occupied areas. Airflow volume (Q) in cfm through inlet and outlet openings of equal area from wind can be estimated by:

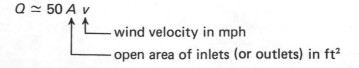

$$Q \simeq 50\,A\,v$$

wind velocity in mph

open area of inlets (or outlets) in ft²

CLIMATE AND SHELTER CONSIDERATIONS—EFFECT OF BUILDING SHAPE AND ORIENTATION ON WIND

Wind protection can be achieved by careful placement of buildings and use of windbreaks. Orientation with respect to prevailing winds is especially important for tall buildings where there will be little shelter from trees and structures.

Building Shape

Moving air generally follows the contour of curved surfaces in a predictable pattern. Sharp corners (and/or rough surfaces) will cause separation and eddies as shown by the examples below.

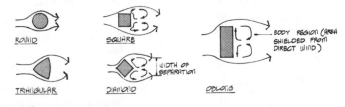

REFERENCE

Evans, B. H., *"Natural Air Flow Around Buildings."* Texas Engineering Experiment Station, Report 59, March 1957, p. 12.

Multiple-Building Layouts

The layouts, shown in plan view at the left and center below, are useful for deflecting winter winds away from downwind buildings. The staggered layout on the right is appropriate for summer breeze distribution to adjacent structures.

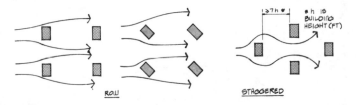

REFERENCE

Olgyay, V., *Design with Climate.* Princeton, New Jersey: Princeton University Press, 1963, p. 101.

Tall Buildings

Tall buildings can deflect wind towards the ground causing high wind speeds (>10 mph) in pedestrian areas. Large podium bases, wide canopies, enclosed malls, etc. can be used to shield pedestrian areas.

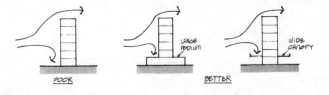

REFERENCE

Penwarden, A. D., "Wind Environment Around Tall Buildings." *Building Research Station Digest,* May 1972.

WIND SPEED

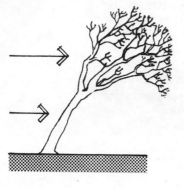

The table below describes some effects of wind on people. Wind speed is expressed in miles per hour (mph). The magnitude, direction, and frequency of wind occurrence at a specific location can be estimated from meteorological data (e.g., available from the National Weather Service). Buildings should be designed so that wind speeds near pedestrian areas and openings will be less than 10 mph. Scale models of tall buildings are often studied in wind tunnels to help predict safe wind speeds.

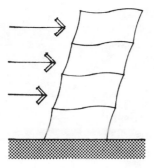

Wind Speed (in mph)*	Typical Effect of Wind on People
0 to 2	No noticeable wind.
2 to 10	Wind felt on face, hair disturbed.
10 to 20	Raises dust, dry soil, and loose papers; hair and clothing disarranged.
20 to 25	Wind force felt on body.
25 to 30	Umbrellas used with difficulty.
30 to 55	Difficult to walk, people can be blown over by gusts.
55 to 100	Hurricane force winds, dangerous to people and structures.
100 and above	Tornado force winds, extremely dangerous to people and structures.

*Wind velocity in fpm is 88 times a given wind speed in mph.

CLIMATE AND SHELTER CONSIDERATIONS—EFFECT OF BARRIERS ON WIND VELOCITY

Shown below are the effects of windbreaks on wind velocity by: thin cottonwood belt, 65 ft high; dense ash belt, 40 ft high; and board fence barrier, 33% solid, 16 ft high. Velocities were recorded at 1 ft-4 in. above the ground for 15 mph winds. The wind velocity, for example, will be reduced 25% by the dense ash belt at a distance of 330 ft behind the windbreak (See dashed lines on chart).

REFERENCE

Stoeckeler, J. H., and Williams, A. R., "Windbreaks and Shelterbelts." *Yearbook of Agriculture,* Washington, DC, 1949.

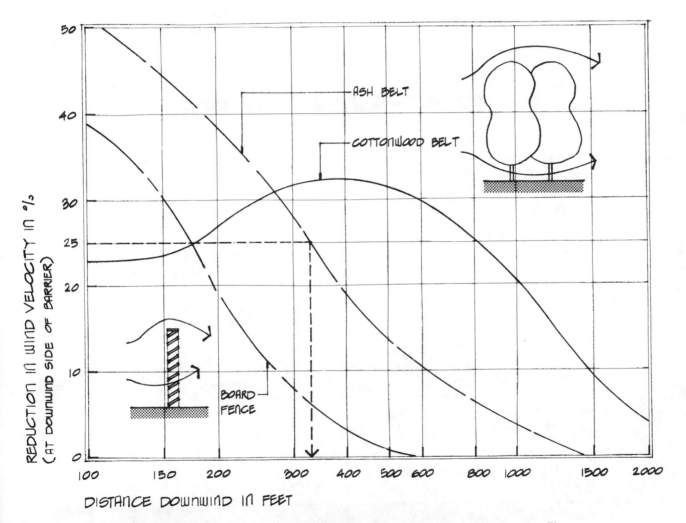

Note: Windbreaks are most effective when placed perpendicular to the prevailing winds. The more open a windbreak (e.g., horizontal louvers or vertical laths), the greater the downwind distance of effectiveness as a larger downwind (or "leeward") side wake is formed.

CLIMATE AND SHELTER CONSIDERATIONS—WIND CONTROL BARRIERS

Examples of wind control by tree and plant barriers are given below. It should be remembered that these natural elements do *not* always have a predictable size, shape, growth rate or site longevity.

Tree-Hedge Barrier

A tree-hedge barrier combination can be used to control wind direction.

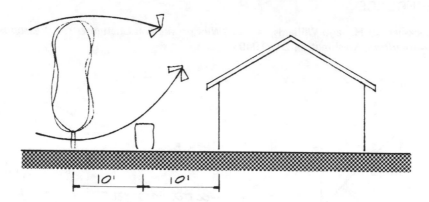

Note: With the hedge and tree positions reversed, airflow would be directed towards the wall rather than over the building.

Windbreak (Shown for Evergreens)

A windbreak of trees can reduce air velocity by about 35% over a horizontal distance of 5H (where H is the windbreak height). It can also reduce air velocity by about 25% over distances as great as 10H. Note that blocks of planting, having a mixture of tree sizes, in general are more effective than narrow bands. For open-field windbreaks, the most efficient length to width ratio is usually about 12 to 1.

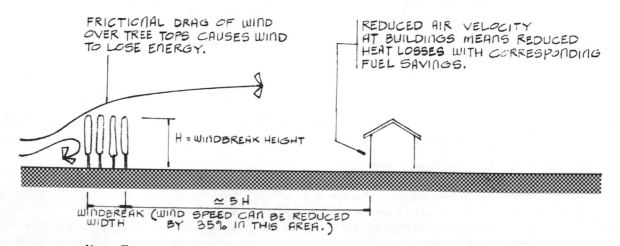

Note: Trees and vegetation can also be used to control drifting snow as well as reduce heat loss from buildings in winter. In summer the leaves absorb solar radiation and evaporation from them tends to cool the adjacent air.

CLIMATE AND SHELTER CONSIDERATIONS—WIND CONTROL FOR VENTILATION

Trees and vegetation can be used to direct beneficial air motion into buildings. Be careful to avoid locating trees and vegetation where they might eliminate desirable cooling breezes during the overheated periods. Nevertheless, mechanical air-conditioning will be required for most situations as outdoor breeze control and air cleaning are not possible with natural ventilation.

A. Tree 5 Feet from Building

Note: Shaded area indicates extent of beneficial air motion.

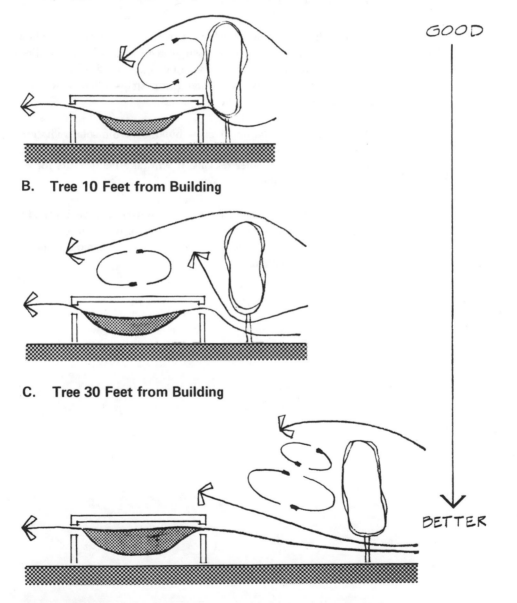

B. Tree 10 Feet from Building

C. Tree 30 Feet from Building

REFERENCE

White, R. F., "Effects of Landscape Development on Natural Ventilation of Buildings and Their Adjacent Area". Texas Engineering Experiment Station, Report 45, March 1945.

CHECKLIST FOR CLIMATE AND SHELTER CONSIDERATIONS

Analysis of temperature and relative humidity data for a climate region will determine if moisture, winds, sunshine, or shading is needed. Try to achieve comfort conditions by natural means. For example, in summer prevailing winds and breezes can be used to help cool buildings by removing heat and moisture. In winter, trees, vegetation, and earth berms can be used to shield buildings from prevailing winds.

Buildings with a minimum of perimeter surface area (e.g., spherical or cylindrical shapes) will have smaller heating and cooling loads than oblong shapes of equal floor area.

Orient oblong buildings so the long axis is in an east-west direction (short sides on the east and west).

Solar radiation should be reduced before it can penetrate the building exterior. For every 100 sq ft of unshaded, clear glass eliminated from building exterior walls, the cooling load will be reduced by about one ton of refrigeration. Glazing can often be reduced to narrow horizontal vision strips where extensive exterior views are unnecessary.

Use trees and vegetation to shade buildings from solar radiation. However, do *not* include beneficial shading effects of trees in heat gain computations as trees can die or be removed for other reasons. It will take many years for replacement trees to achieve equal shading effectiveness.

Overhangs, louvers, panels, and egg-crate devices can be designed to provide effective shading. Overhangs are particularly effective on south exposures. In tall buildings, balconies, or upper floors (e.g., upside-down pyramidal shapes), can be designed to provide effective shading.

BUILDING
MATERIALS

BUILDING MATERIALS—TRADITIONAL STRUCTURES

The examples below show how some traditional shelters effectively coped with widely varying climatic demands.

Cool Region

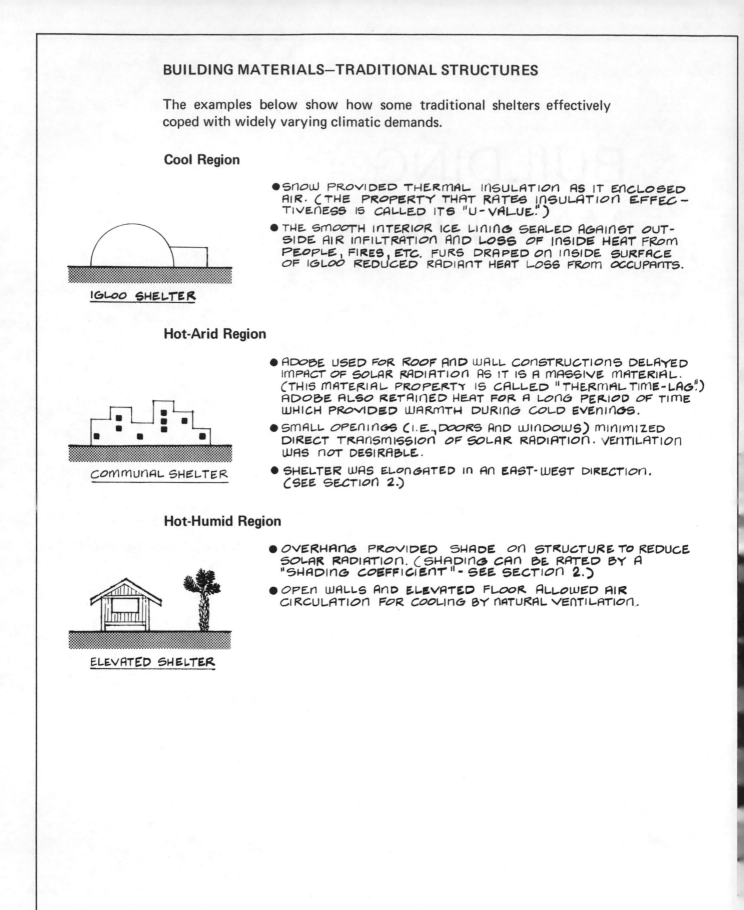

IGLOO SHELTER

- SNOW PROVIDED THERMAL INSULATION AS IT ENCLOSED AIR. (THE PROPERTY THAT RATES INSULATION EFFECTIVENESS IS CALLED ITS "U-VALUE.")
- THE SMOOTH INTERIOR ICE LINING SEALED AGAINST OUTSIDE AIR INFILTRATION AND LOSS OF INSIDE HEAT FROM PEOPLE, FIRES, ETC. FURS DRAPED ON INSIDE SURFACE OF IGLOO REDUCED RADIANT HEAT LOSS FROM OCCUPANTS.

Hot-Arid Region

COMMUNAL SHELTER

- ADOBE USED FOR ROOF AND WALL CONSTRUCTIONS DELAYED IMPACT OF SOLAR RADIATION AS IT IS A MASSIVE MATERIAL. (THIS MATERIAL PROPERTY IS CALLED "THERMAL TIME-LAG.") ADOBE ALSO RETAINED HEAT FOR A LONG PERIOD OF TIME WHICH PROVIDED WARMTH DURING COLD EVENINGS.
- SMALL OPENINGS (I.E., DOORS AND WINDOWS) MINIMIZED DIRECT TRANSMISSION OF SOLAR RADIATION. VENTILATION WAS NOT DESIRABLE.
- SHELTER WAS ELONGATED IN AN EAST-WEST DIRECTION. (SEE SECTION 2.)

Hot-Humid Region

ELEVATED SHELTER

- OVERHANG PROVIDED SHADE ON STRUCTURE TO REDUCE SOLAR RADIATION. (SHADING CAN BE RATED BY A "SHADING COEFFICIENT"- SEE SECTION 2.)
- OPEN WALLS AND ELEVATED FLOOR ALLOWED AIR CIRCULATION FOR COOLING BY NATURAL VENTILATION.

BUILDING MATERIALS—CONDUCTION, CONVECTION AND RADIATION

Shown below are examples of heat flow through building constructions. The arrows indicate the direction of heat flow by conduction, convection, or radiation.

Conduction

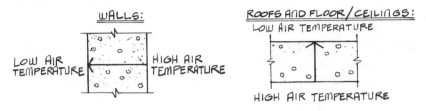

Conduction is the heat transfer through solid materials from the warmer to cooler particles. A familiar example of conduction would be a cold metal poker held in a hot fire. The transfer of heat from the fire to the opposite end is by conduction (i.e., heated particles "bump" adjacent particles and pass their increased thermal energy to them).

Convection

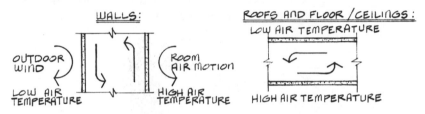

Convection is the heat transfer in air by the motion of heated air from a warmer to cooler surface. An example of convection is cigarette smoke flow toward room lamps—the heated air near the lamp rises because it is less dense.

Radiation

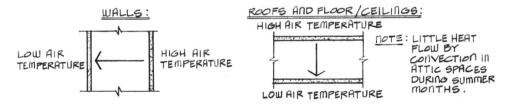

Radiation is the heat transfer by electromagnetic waves from a warmer to cooler surface. An example of radiation is the sun warming a car's occupants on a cold day. The radiant heat is transferred directly and is not affected by the cool temperature of the surrounding air.

Note: Building heat loss (or gain) can be reduced by the use of insulating materials having slow conduction rates, by air spaces, and by reflective linings (e.g., bright aluminum foil) in air spaces to reduce radiant heat transfer. Often insulating materials are manufactured with an adhered layer of metallic foil. When used in wall or roof cavities these materials will provide all three kinds of heat flow reduction.

BUILDING MATERIALS—UNITS FOR HEAT FLOW

Conductivity (k) is the rate of heat flow in Btuh through 1 sq ft of a homogeneous material, 1 in. thick, per 1°F temperature drop through the material. Conductance (C) is also a rate of heat flow through 1 sq ft per 1°F temperature drop; however, conductance values are usually stated for a specific thickness *not* per 1 in. For homogeneous materials (e.g., concrete, fibrous insulation, plaster), C equals k/x where x is the material thickness in inches. Heat flow coefficients are derived from laboratory tests.

The resistance (R) is the reciprocal of the heat flow coefficients ($1/C$, x/k, etc.). The overall resistance for a construction is equal to the sum of its component resistances. Consequently, when the resistances of a construction are known, the overall or total resistance (R_T) and its reciprocal, the U–coefficient of heat transmission, can be found by the formula:

$$U = \frac{1}{R_T}$$

total resistance in hrs/Btu/ft²/°F

overall coefficient of heat transmission (or "U-value") in Btuh/ft²/°F

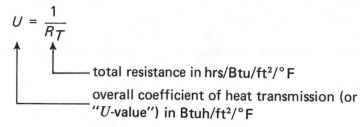

For 1°F temperature drop through 1 sq ft of above insulation (glass fiberboard) materials:

Conductivity (k)	0.25	0.25	0.25	Btuh/ft²/in./°F
Conductance (C)	0.25	$\frac{0.25}{2} = 0.13$	$\frac{0.25}{4} = 0.06$	Btuh/ft²/°F
Resistance (R)	4.0	8.0	16.0	hrs/Btu/ft²/°F

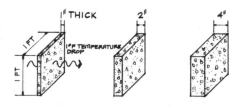

For 1°F temperature drop through 1 sq ft of above concrete (sand and gravel) materials:

Conductivity (k)	12.0	12.0	12.0	Btuh/ft²/in./°F
Conductance (C)	12.0	$\frac{12}{2} = 6.0$	$\frac{12}{4} = 3.0$	Btuh/ft²/°F
Resistance (R)	0.08	0.17	0.33	hrs/Btu/ft²/°F

BUILDING MATERIALS—HEAT FLOW COEFFICIENTS

Conventional symbols and units for the thermal properties of homogeneous materials, nonhomogeneous materials (e.g., hollow tile, concrete block, gypsum lath), and air are listed below. The thermal property letter symbols also are shown on the wall section to indicate where they apply. Note that temperature gradients, which show the temperature drop in °F through a construction, can be computed from thermal properties (See page 57).

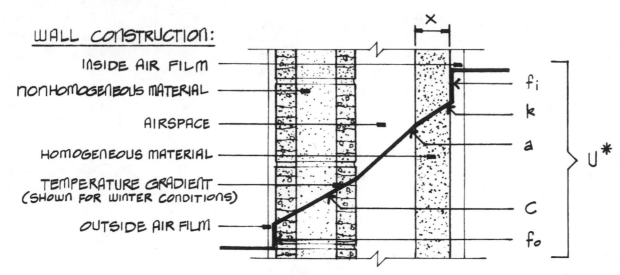

WALL CONSTRUCTION:
- INSIDE AIR FILM
- NONHOMOGENEOUS MATERIAL
- AIRSPACE
- HOMOGENEOUS MATERIAL
- TEMPERATURE GRADIENT (SHOWN FOR WINTER CONDITIONS)
- OUTSIDE AIR FILM

* SYMBOLS LISTED TO INDICATE WHERE THEY APPLY. UNITS ARE GIVEN IN THE TABLE.

Symbol	Thermal Property	Units	Resistance
f_o	Outside air film conductance	Btuh/ft²/°F	$1/f_o$
C	Conductance	Btuh/ft²/°F	$1/C$
a	Air space conductance	Btuh/ft²/°F	$1/a$
k	Conductivity	Btuh/ft²/in./°F	x/k
f_i	Inside air film conductance	Btuh/ft²/°F	$1/f_i$
R	Resistance	hrs/Btu/ft²/°F	R
U	U-coefficient of heat transmission (or U-value)	Btuh/ft²/°F	$1/U$

U-coefficient formula:

$$U = \frac{1}{R_T} = \frac{1}{\left(\dfrac{1}{f_i} + \dfrac{x}{k} + \dfrac{1}{a} + \dfrac{1}{C} + \dfrac{1}{f_o}\right)}$$

HEAT FLOW COEFFICIENTS FOR COMMON BUILDING MATERIALS

Conductance (f_i, f_o, a, and C) and conductivity (k) values are in Btuh per sq ft per °F temperature difference. Resistance (R) values are the reciprocals of the given conductance.

Air Films

Position of Surface Inside (or Still) Air:	Heat Flow*	Aluminum-coated Paper		Ordinary Building Materials**	
		f_i	R	f_i	R
Horizontal	Up	0.91	1.10	1.63	0.61
Slope, 45°	Up	0.88	1.14	1.60	0.62
Vertical	Horizontal	0.74	1.35	1.46	0.68
Slope, 45°	Down	0.60	1.67	1.32	0.76
Horizontal	Down	0.37	2.70	1.08	0.92

Any Position for Outside (or Moving) Air:				f_o	R
15 mph wind	Any (for winter computations)			6.00	0.17
7½ mph wind	Any (for summer computations)			4.00	0.25

Air Spaces

Position of Air Space	Heat Flow*	Thickness (in.)	Temperature Conditions	Aluminum-coated Paper		Ordinary Building Materials**	
				a	R	a	R
Horizontal	Up	¾	Winter	0.59	1.71	1.15	0.87
		¾	Summer	0.61	1.63	1.32	0.76
		4	Winter	0.50	1.99	1.06	0.94
		4	Summer	0.53	1.87	1.25	0.80
Slope, 45°	Up	¾	Winter	0.50	2.02	1.06	0.94
		¾	Summer	0.53	1.90	1.24	0.81
		4	Winter	0.47	2.13	1.04	0.96
		4	Summer	0.51	1.98	1.22	0.82
Vertical	Horizontal	¾	Winter	0.42	2.36	0.99	1.01
		¾	Summer	0.48	2.10	1.19	0.84
		4	Winter	0.43	2.34	0.99	1.01
		4	Summer	0.46	2.16	1.10	0.91
Horizontal	Down	¾	Winter	0.42	2.39	0.98	1.02
		¾	Summer	0.48	2.08	1.19	0.84
		1½	Winter	0.31	3.21	0.88	1.14
		1½	Summer	0.36	2.76	1.07	0.93
		4	Winter	0.25	4.02	0.81	1.23
		4	Summer	0.30	3.38	1.01	0.99
Slope, 45°	Down	¾	Winter	0.42	2.40	0.98	1.02
		¾	Summer	0.48	2.09	1.19	0.84
		4	Winter	0.36	2.75	0.93	1.08
		4	Summer	0.40	2.50	1.11	0.90

*For ceiling constructions, the direction of heat flow normally will be "up" in winter, "down" in summer. For floor constructions, the direction of heat flow will be "down" in winter, "up" in summer. Heat flow through walls is in a "horizontal" direction.

**Wood, paper, glass, masonry, nonmetallic paints, etc. are considered "ordinary building materials."

HEAT FLOW COEFFICIENTS (Continued)

Building Materials

Building Material	Thickness (in.)	Conductivity k (for 1-in. thickness)	Conductance C (for given thickness)	Resistivity 1/k (per 1-in. thickness)	Resistance 1/C or R (for given thickness)
Insulating:					
Cane fiberboard	$\frac{1}{2}$		0.80		1.25
	$\frac{3}{4}$		0.53		1.89
Cellular glass		0.40		2.50	
Cellulose fibers		0.27		3.70	
Corkboard, 8 pcf		0.27		3.70	
Expanded polystyrene:					
extruded, 3.5 pcf		0.19		5.26	
molded beads, 1 pcf		0.28		3.57	
Expanded polyurethane, 2.5 pcf		0.16		6.25	
Glass-fiber		0.25		4.00	
Mineral fiber batt	$2\frac{3}{4}$		0.14		7.00
	$3\frac{1}{2}$		0.09		11.00
	$6\frac{1}{2}$		0.05		19.00
Mineral fiberboard, 18 pcf		0.35		2.86	
Mineral (or glass-fiber) wool		0.27		3.70	
Shredded wood fiberboard	2		0.28		3.50
	$2\frac{1}{2}$		0.23		4.38
	3		0.19		5.25
Styrofoam, 2.3 pcf	$\frac{3}{4}$	0.20	0.24	5.00	4.17
Vermiculite, 4 to 6 pcf		0.44		2.27	
Exterior:					
Asbestos-cement shingles			4.75		0.21
Asphalt roll roofing (or siding)			6.50		0.15
Asphalt sheathing	$\frac{1}{2}$		0.69		1.46
Asphalt shingles			2.27		0.44
Built-up roofing, T & G	$\frac{3}{8}$		3.00		0.33
Gypsum sheathing	$\frac{3}{8}$		3.10		0.32
	$\frac{5}{8}$		1.75		0.57
Hardwood: maple, oak, etc.		1.10		0.91	
Insulating fiberboard	$\frac{25}{32}$		0.49		2.06
Lapped siding, plywood	$\frac{3}{8}$	0.80	1.59	1.25	0.59
Slate	$\frac{1}{2}$	10.00	20.00	0.10	0.05
Softwood: fir, yellow pine, etc.		0.80		1.25	
Stucco		5.00		0.20	
Wood shingles			1.06		0.94
Interior:					
Carpet on fibrous underlay			0.48		2.08
Carpet on foam rubber underlay			0.81		1.23
Cement plaster (sand aggregate)	$\frac{3}{8}$	5.00	13.33	0.20	0.08
	$\frac{3}{4}$	5.00	6.66	0.20	0.15
Cork tile	$\frac{1}{8}$		3.60		0.28

HEAT FLOW COEFFICIENTS (Continued)

Building Materials (Continued)

Building Material	Thickness (in.)	Conductivity k (for 1-in. thickness)	Conductance C (for given thickness)	Resistivity 1/k (per 1-in. thickness)	Resistance 1/C or R (for given thickness)
Interior:					
Gypsum board	$\frac{3}{8}$		3.10		0.32
	$\frac{1}{2}$		2.25		0.45
	$\frac{5}{8}$		1.75		0.57
Gypsum lath and plaster	$\frac{7}{8}$		2.44		0.41
Gypsum plaster (lightweight aggregate)	$\frac{1}{2}$		3.12		0.32
	$\frac{5}{8}$		2.67		0.39
Gypsum plaster (sand aggregate)	$\frac{1}{2}$		11.10		0.09
	$\frac{5}{8}$		9.10		0.11
Metal lath and plaster (sand aggregate)	$\frac{3}{4}$		7.70		0.13
Metal lath and plaster (lightweight aggregate)	$\frac{3}{4}$		2.13		0.47
Plywood	$\frac{3}{8}$	0.80	2.13	1.25	0.47
Tile: asphalt, linoleum, vinyl, or rubber			20.00		0.05
Wood subfloor	$\frac{3}{4}$		1.06		0.94
Masonry:					
Brick, common, 120 pcf	4	5.00	1.25	0.20	0.80
Brick, face, 130 pcf	4	9.00	2.27	0.11	0.44
Cement mortar		5.00		0.20	
Clay tile, hollow	3		1.25		0.80
	4		0.90		1.11
	6		0.66		1.52
	8		0.54		1.85
	10		0.45		2.22
	12		0.40		2.50
Concrete:					
sand and gravel aggregate, 140 pcf		12.00		0.08	
lightweight aggregate, 60 pcf		1.70		0.59	
30 pcf		0.90		1.11	
20 pcf		0.70		1.43	
Concrete block, hollow, three cells:					
cinder aggregate	4		0.90		1.11
	8		0.58		1.72
	12		0.53		1.89
lightweight aggregate	4		0.67		1.50
	8		0.50		2.00
	12		0.44		2.27
sand and gravel aggregate	4		1.40		0.71
	8		0.90		1.11
	12		0.78		1.28
Gypsum partition tile, hollow	4		0.60		1.67

HEAT FLOW COEFFICIENTS (Continued)

Building Materials (Continued)

Building Material	Thickness (in.)	Conductivity k (for 1-in. thickness)	Conductance C (for given thickness)	Resistivity 1/k (per 1-in. thickness)	Resistance 1/C or R (for given thickness)
Masonry:					
Stone, lime, or sand		12.50		0.08	
Stucco		5.00		0.20	
Terrazzo	1	12.50	12.50	0.08	0.08
Miscellaneous:					
Aluminum		1536.00		negligible	
Asphalt 15-lb felt, two layers			8.35		0.12
Corrugated paper (3.7 pcf)		0.45		2.13	
Glass, average		6.00		0.17	
Lead		240.00		negligible	
Polyethylene sheet			negligible		negligible
Sheet metal		456.00		negligible	
Soil, dry, loose		3.00		0.33	
Soil, average		7.00		0.14	
Soil, damp, packed		21.00		0.05	
Steel		312.00		negligible	
Water		5.50		0.18	

Source: *ASHRAE Guide and Data Book: Fundamentals* (N.Y.: American Society of Heating, Refrigerating and Air-Conditioning Engineers, Inc. (ASHRAE), 1972). This book is published periodically by ASHRAE, Inc., 345 East 47th Street, New York, N Y 10017.

TEST REFERENCE

"Thermal Conductivity of Materials by Means of the Guarded Hot Plate," ASTM Method C177-63. American Society for Testing and Materials (ASTM), 1916 Race Street, Philadelphia, PA 19103.

BUILDING MATERIALS—EXAMPLE PROBLEMS: *U*-COEFFICIENTS FOR WALLS

Wall Construction "A-1"

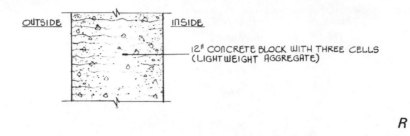

		R
1.	Inside air	0.68
2.	12 in. concrete block	2.27
3.	Outside air (moving at 15 mph)	0.17
	Total *R* :	3.12*

$$U = \frac{1}{R_T} = \frac{1}{3.12} = \boxed{0.32 \text{ Btuh/ft}^2/°\text{F}}$$

*Decimal totals are used only to check computations, as fractional Btu's will be of little significance. Accordingly, *U*-value answers should be rounded to the nearest two decimal places.

Wall Construction "A-2"

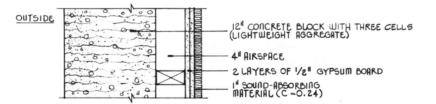

		R
1.	Inside air	0.68
2.	1 in. sound-absorbing material (*C* = 0.24)	$\frac{1}{0.24}$ = 4.17
3.	2 layers ½ in. gypsum board	2 X 0.45 = 0.90
4.	4 in. air space (*Note:* both surfaces "ordinary building materials")	1.01
5.	12 in. concrete block	2.27
6.	Outside air	0.17
	Total *R* :	9.20

$$U = \frac{1}{R_T} = \frac{1}{9.20} = \boxed{0.11 \text{ Btuh/ft}^2/°\text{F}}$$

Note: *U*-value example problems in this book for walls and roofs are computed for winter conditions. For summer, use outside air surface resistance $(1/f_o)$ of 0.25 and the appropriate inside air surface and air space resistances from the tables.

BUILDING MATERIALS—EXAMPLE PROBLEMS: *U*-COEFFICIENTS FOR ROOFS

Roof Construction "B-1"

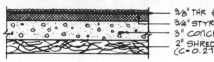

3/8" TAR & GRAVEL (BUILT-UP ROOFING)
3/4" STYROFOAM INSULATION (2.3 PCF)
3" CONCRETE SLAB (SAND AND GRAVEL)
2" SHREDDED WOOD FIBERBOARD (C=0.27)

			R
1.	Outside air		0.17
2.	$\frac{3}{8}$ in. tar & gravel		0.33
3.	$\frac{3}{4}$ in. styrofoam insulation (2.3 pcf)		4.17
4.	3 in. concrete slab	3 X 0.08 =	0.24
5.	2 in. shredded wood fiberboard	$\frac{1}{0.27}$ =	3.70
6.	Inside air		0.61
		Total *R* :	9.22

$$U = \frac{1}{R_T} = \frac{1}{9.22} = \boxed{0.11 \text{ Btuh/ft}^2/^\circ\text{F}}$$

Roof Construction "B-2"

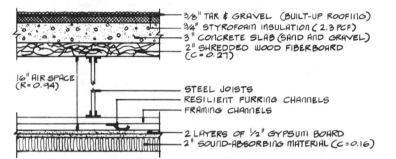

3/8" TAR & GRAVEL (BUILT-UP ROOFING)
3/4" STYROFOAM INSULATION (2.3 PCF)
3" CONCRETE SLAB (SAND AND GRAVEL)
2" SHREDDED WOOD FIBERBOARD (C=0.27)

16" AIR SPACE (R=0.94)

STEEL JOISTS
RESILIENT FURRING CHANNELS
FRAMING CHANNELS

2 LAYERS OF 1/2" GYPSUM BOARD
2" SOUND-ABSORBING MATERIAL (C=0.16)

			R
1.	Outside air		0.17
2.	$\frac{3}{8}$ in. tar & gravel		0.33
3.	$\frac{3}{4}$ in. styrofoam insulation (2.3 pcf)		4.17
4.	3 in. concrete slab		0.24
5.	2 in. shredded wood fiberboard		3.70
6.	16 in. air space (*Note:* use maximum thickness value from table)		0.94
7.	Steel joists and channels		0.00
8.	2 layers $\frac{1}{2}$ in. gypsum board	2 X 0.45 =	0.90
9.	2 in. sound– absorbing material	$\frac{1}{0.16}$ =	6.25
10.	Inside air		0.61
		Total *R* :	17.31

$$U = \frac{1}{R_T} = \frac{1}{17.31} = \boxed{0.06 \text{ Btuh/ft}^2/^\circ\text{F}}$$

BUILDING MATERIALS—SUGGESTED *U*-COEFFICIENTS FOR WALLS

Human thermal comfort in winter can be affected by the radiation of body heat to cold indoor surfaces (floor, walls, or ceiling) and conversely, in summer, by the body heat gain from warm surfaces. However, in winter, surface temperatures about 5°F (or less) below indoor air temperature will generally provide satisfactory radiant comfort conditions. The table below shows wall *U*-values needed in four climate regions for a given wall surface temperature requirement. For example, the walls of a motel, located in the temperate region, with an indoor design temperature of 75°F will require a *U*-value of 0.10 Btuh/ft²/°F or less. (Enter table at 70°F row (i.e., 75-5°F) and read *U*-value at intersection with climate column.) For energy conservation, however, a *U*-value of 0.06 (or less) is often preferred in the cool and temperate climate regions.

REFERENCE

Rogers, T. S., *Thermal Design of Buildings.* New York: John Wiley & Sons, Inc., 1964.

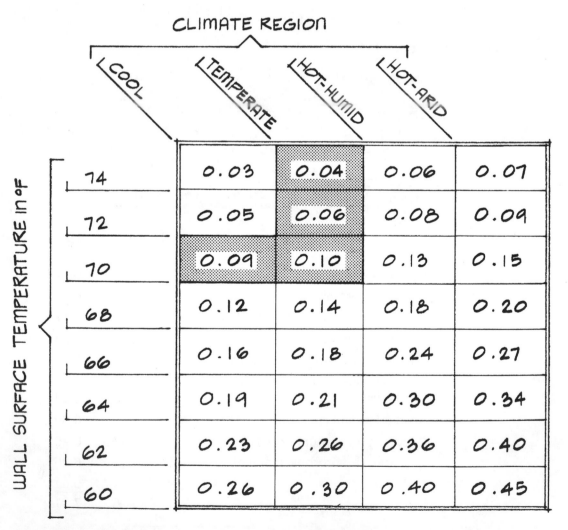

CLIMATE REGION

WALL SURFACE TEMPERATURE in °F	COOL	TEMPERATE	HOT-HUMID	HOT-ARID
74	0.03	0.04	0.06	0.07
72	0.05	0.06	0.08	0.09
70	0.09	0.10	0.13	0.15
68	0.12	0.14	0.18	0.20
66	0.16	0.18	0.24	0.27
64	0.19	0.21	0.30	0.34
62	0.23	0.26	0.36	0.40
60	0.26	0.30	0.40	0.45

Note: For ceilings, decrease the *U*-values in the table by 15%. This table is also valid for summer since the *U*-values that are best in winter are also best in summer.

BUILDING MATERIALS—TEMPERATURE GRADIENT

The temperature gradient shows the temperature of inside building surfaces and the location within exterior constructions where condensation is most likely to occur. The gradient can be found by proportioning the overall temperature drop through a construction by the ratios of the component resistances to the total resistance value. Example computations for 12 in. concrete block wall with an inside surface of 1 in. gypsum board is given below with and without insulation.

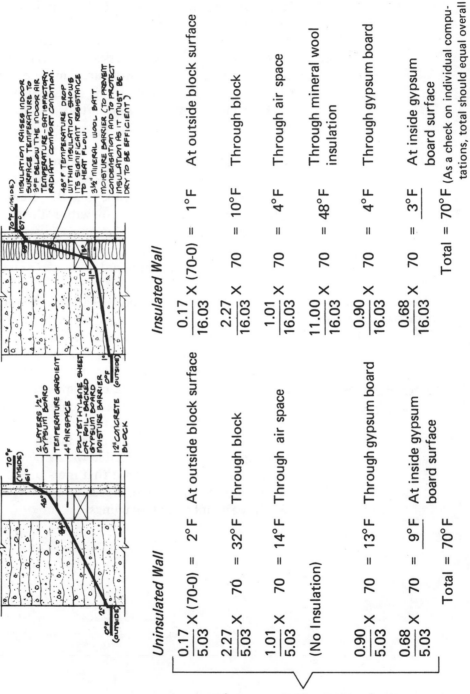

Temperature Gradient Analysis

Uninsulated Wall

$$\frac{0.17}{5.03} \times (70-0) = 2°F \quad \text{At outside block surface}$$

$$\frac{2.27}{5.03} \times 70 = 32°F \quad \text{Through block}$$

$$\frac{1.01}{5.03} \times 70 = 14°F \quad \text{Through air space}$$

(No Insulation)

$$\frac{0.90}{5.03} \times 70 = 13°F \quad \text{Through gypsum board}$$

$$\frac{0.68}{5.03} \times 70 = \underline{9°F} \quad \text{At inside gypsum board surface}$$

$$\text{Total} = 70°F$$

Insulated Wall

$$\frac{0.17}{16.03} \times (70-0) = 1°F \quad \text{At outside block surface}$$

$$\frac{2.27}{16.03} \times 70 = 10°F \quad \text{Through block}$$

$$\frac{1.01}{16.03} \times 70 = 4°F \quad \text{Through air space}$$

$$\frac{11.00}{16.03} \times 70 = 48°F \quad \text{Through mineral wool insulation}$$

$$\frac{0.90}{16.03} \times 70 = 4°F \quad \text{Through gypsum board}$$

$$\frac{0.68}{16.03} \times 70 = \underline{3°F} \quad \text{At inside gypsum board surface}$$

$$\text{Total} = 70°F \quad \text{(As a check on individual computations, total should equal overall temperature drop.)}$$

MOISTURE BARRIERS

Condensation will occur when the temperature of a surface falls below the "dew-point" of the surrounding air. In air-conditioning, the dew-point is the temperature at which water vapor will be extracted from air (e.g., condensation on the outside of an iced tea glass on a hot day). In nature when warm, moist air is cooled or contacts a cold surface, the moisture condenses as dew or fog. Note that the greater the moisture content of air (usually expressed as % relative humidity), the higher the dew-point temperature. Moisture can be removed from kitchens, bathrooms, etc. by exhaust fans. In winter, it is important that exhausted air be replaced by cold, dry outside air to lower the indoor relative humidity.

At a typical indoor design temperature of 70°F, and assuming that 30% relative humidity will not be exceeded for any extended time, the dew-point will be 37°F (See Appendix B: Dew-point Table). Therefore, hidden condensation will occur within the air space of the example uninsulated construction on the preceding page when the room relative humidity is 30% or greater unless a moisture barrier is used on the air space side of the gypsum board.

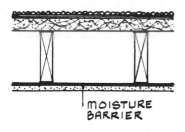

MOISTURE
BARRIER

Typical moisture (or "vapor") barrier materials are polyethylene sheet, insulation back-up paper, etc. Moisture barriers should always be installed on the warm side of insulation materials. The closer the moisture barrier is to the sources of moisture (See example roof construction at left), the greater will be the tolerable moisture content of air in a given space. In general, for cool and temperate climate regions place the moisture barrier near the inner surface of exterior constructions, for hot-humid regions near the outside surface. An acceptable material is normally rated at 0.1 perm (preferably less) for most building applications. Many thermal insulating materials are available with a factory applied moisture barrier covering. It is very important that moisture barriers be installed without any holes or cracks. Where pipes and conduit must penetrate the barrier, carefully fit and seal all openings.

MOISTURE BARRIER RATINGS

Moisture barrier ratings in perms are listed below for design guidance only. The values given are for comparison purposes as actual moisture barrier selection should be based on data from laboratory tests according to current ASTM standards. The unit "perm" indicates a water vapor transmission rate of 1 grain/ft²/hr/vapor pressure difference between the two sides of the material (1 grain is equivalent to about one drop of water).

Material	Permeance (in perms)
Materials used in construction:	
1. Concrete, 4 in. thick (150 pcf)	12.0 - 20.0
2. Concrete block, 8 in., with cells, limestone aggregate	2.4
3. Plaster on metal lath ($\frac{3}{4}$ in.)	15.0
4. Plaster on gypsum lath ($\frac{3}{4}$ in.)	20.0
5. Plywood, douglas-fir, interior, $\frac{1}{4}$ in.	1.9
6. Roofing, built-up, hot mopped	0.0
Thermal insulations:	
7. Mineral wool, unprotected	29.0
Plastic and metal foils:	
8. Aluminum foil (1 mil)	0.0
9. Polyethylene (4 mil)	0.08
10. Polyethylene (6 mil)	0.06
11. Polyethylene (8 mil)	0.04
Building papers, felts, and roofing papers:	
12. Kraft paper and asphalt laminated, reinforced 30-120-30 (34 p500sf)*	1.8
13. Saturated and coated roll roofing (326 p500sf)	0.24
14. 15-lb asphalt felt (70 p500sf)	5.6
15. 15-lb tar felt (70 p500sf)	18.2
Liquid-applied coating materials:	
16. Enamel paint on smooth plaster (two coats)	0.5 - 1.5
17. Various primers (two coats) plus one coat flat oil-base paint on plaster	1.6 - 3.0
Miscellaneous:	
18. Expanded polystyrene, molded (1 in.)	0.5 - 1.0

*"p500sf" is surface weight unit of pounds per 500 sq ft of material.

Source: *ASHRAE Handbook of Fundamentals* (1972).

TEST REFERENCE

"Water Vapor Transmission of Materials in Sheet Form," ASTM Method E96-66.

BUILDING MATERIALS—CONDENSATION CONTROL BY VENTILATION

It is good practice to provide attic and crawl space ventilation. In winter months, this will allow unwanted moisture to escape before it condenses and will help achieve thermal comfort conditions in the summer. At least two vents are required—preferably at opposite ends of the structure.

Attics

Well-distributed attic soffit, gable, or ridge vents (a combination of soffit inlets and ridge outlets is best) should have at least a total open or "free" area $\frac{1}{300}$ of the ceiling area per FHA recommendations. If openings are covered with bird screens having a mesh size less than $\frac{1}{8}$ in., use $\frac{1}{150}$. Properly vented attic spaces can reduce heat gain through the roof by 20% or more. (To estimate U-value for attic constructions, see formula A-8, page 179.)

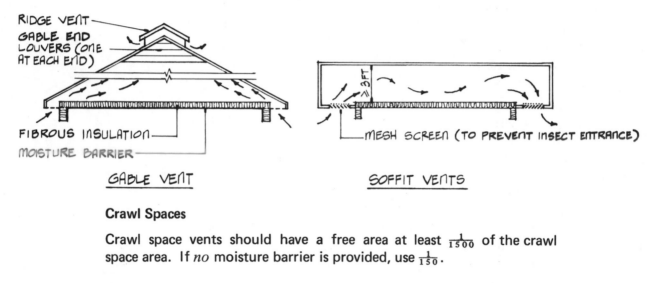

GABLE VENT   SOFFIT VENTS

Crawl Spaces

Crawl space vents should have a free area at least $\frac{1}{1500}$ of the crawl space area. If *no* moisture barrier is provided, use $\frac{1}{150}$.

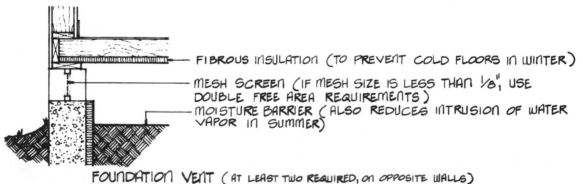

Note: Exhaust fans (or blowers) can be used where natural ventilation is impractical due to nearby structures, design considerations, etc. For preliminary estimates, use a ventilation rate of 1.5 cfm/sq ft of ceiling area to size attic fan units.

BUILDING MATERIALS—CONDENSATION ON GLASS SURFACES IN BUILDINGS

The graph below shows the combination of room relative humidity and outdoor air temperature that will cause condensation on the inside surface of glass at an indoor temperature of 70°F. For example, condensation will occur on glass surfaces for outdoor temperatures of 42°F (or lower) when the room relative humidity is 50%. (See dashed lines on graph.)

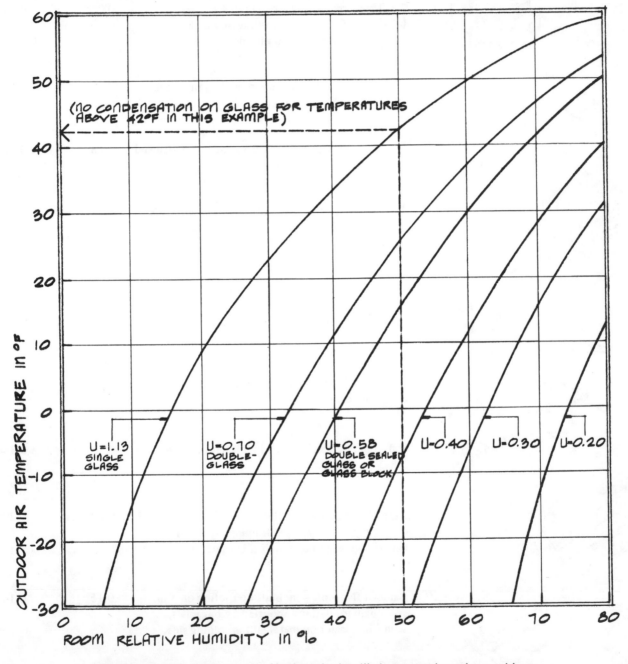

Note: In general, the lower the *U*-value, the less likely are condensation problems. In the hot-humid region, condensation can occur on the outside of glass surfaces. Consequently, use double- or triple-glass where especially cool indoor temperatures are required.

BUILDING MATERIALS – U-VALUES FOR COMMON BUILDING CONSTRUCTIONS

EXTERIOR WALLS

	BUILDING CONSTRUCTION	U-VALUE (Btuh/ft² /°F)	
		Winter	*Summer*
1.	2 × 4 wood studs 16 in. o.c. with $\frac{1}{2}$ in. × 8 in. lapped wood siding and $\frac{1}{2}$ in. asphalt sheathing on outside, and $\frac{1}{2}$ in. gypsum board on inside.	0.22	0.22
2.	2 × 4 wood studs 16 in. o.c. with 4 in. face brick and $\frac{1}{2}$ in. asphalt sheathing on outside, and $\frac{1}{2}$ in. gypsum board on inside.	0.24	0.25
3.	4 in. face brick, 4 in. common brick, $\frac{3}{4}$ in. airspace, and $\frac{1}{2}$ in. gypsum board on inside.	0.29	0.29
4.	8 in. concrete block, 1 in. airspace, $\frac{1}{2}$ in. foil-backed gypsum board on wood furring strips.	0.18	0.17
5.	4 in. common brick, 4 in. airspace, 4 in. concrete block, with 1 in. airspace, and $\frac{1}{2}$ in. gypsum board on wood furring strips.	0.21	0.22
6.	4 in. face brick, 2 in. airspace, 4 in. face brick.	0.37	0.38

Note: To adjust *U*-values for effect of insulation in airspaces, use graph on page 64.

INTERIOR WALLS

7.	2 × 4 wood studs 16 in. o.c. with $\frac{1}{2}$ in. gypsum board on both sides.	0.29	0.29
8.	8 in. concrete block with $\frac{5}{8}$ in. plaster on both sides.	0.31	0.31

ROOFS

9.	Asphalt shingle pitched roof: $\frac{5}{8}$ in. plywood deck, supported by wood framing, $3\frac{1}{2}$ in. airspace and $\frac{1}{2}$ in. gypsum board on ceiling side.	0.22	0.20
10.	Bar joist flat roof: $\frac{3}{8}$ in. built-up roofing on 1 in. styrofoam insulation supported by bar joists, with $\frac{3}{4}$ in. plaster on metal lath on ceiling side.	0.15	0.14
11.	Concrete slab flat roof: $\frac{3}{8}$ in. built-up roofing on 1 in. styrofoam insulation supported by 4 in. concrete slab, with suspended $\frac{3}{4}$ in. plaster on metal lath on ceiling side.	0.13	0.12

BUILDING MATERIALS (Continued)

BUILDING CONSTRUCTION	U-VALUE (Btuh/ft²/°F)	
	Winter	*Summer*

12. Ribbed metal deck: $\frac{3}{8}$ in. built-up roofing on 1 in. styrofoam insulation supported by ribbed metal deck on bar joists, with $\frac{3}{4}$ in. plaster on metal lath on ceiling side. — 0.14 / 0.13

ROOFS (CONT.)

Note: To adjust *U*-values for effect of insulation above ceilings, use graph on page 64. For example, if 4 in. glass-fiber insulation is installed above the ceiling of construction number 12, the improved winter *U*-value will be 0.05.

FLOORS

13. 2 X 10 wood joists 16 in. o.c. with hardwood and wood subfloor on floor side, and $\frac{3}{8}$ in. gypsum board on other side. — 0.27 / 0.29

Note: For concrete slabs on grade, see page 74.

DOORS

14. $1\frac{3}{4}$ in. solid wood core door with weather-stripping. — 0.46 / 0.45

15. Construction number 14 with wood storm door. — 0.26 / —

16. $1\frac{3}{4}$ in. hollow 16 gage steel door, glass-fiber filled core with weatherstripping. — 0.59 / 0.58

GLAZING

17. $\frac{1}{8}$ in. (to $\frac{1}{2}$ in.) thick single plate-glass pane. — 1.13 / 1.06

18. Double-glass: two panes separated by $\frac{1}{4}$ in. airspace. — 0.65 / 0.61

19. Construction number 18 with $\frac{1}{2}$ in. airspace. — 0.58 / 0.56

20. Triple-glass: three panes separated by $\frac{1}{4}$ in. airspaces. — 0.47 / 0.45

21. Construction number 20 with $\frac{1}{2}$ in. airspaces. — 0.36 / 0.35

22. $\frac{1}{8}$ in. (to $\frac{1}{4}$ in.) thick single acrylic plastic sheet. — 1.01 / 0.93

SOURCES: 1. *ASHRAE Handbook of Fundamentals* (1972), N.Y.: American Society of Heating, Refrigerating and Air-Conditioning Engineers.
2. Rogers, T.S., *Thermal Design of Buildings*, New York: John Wiley & Sons, Inc., 1964.

BUILDING MATERIALS—*U*-VALUE IMPROVEMENTS FROM ADDED INSULATION

The graph below shows the *U*-value resulting from the addition of insulation to building constructions. For example, if an uninsulated construction has a *U*-value of 0.40, the addition of 2 in. thick insulation will result in an improved U-value of 0.10 Btuh/ft²/°F. (See dashed lines on graph.)

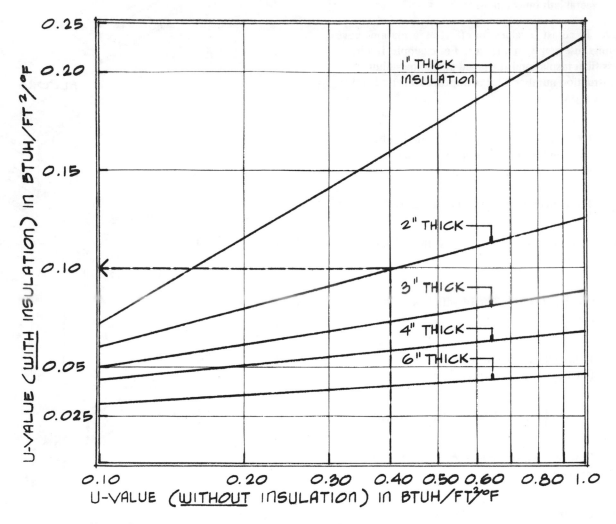

Note: Graph based on a typical conductivity (*k*) value of 0.27 for fibrous ("fuzz") or foamed plastic insulation. If the insulation fills the cavity air space, the *U*-value must be further adjusted for the loss of air space resistance.

A USEFUL RECIPROCAL TABLE

Reciprocals of numbers are often required in heat flow computations. Be careful as heat flow coefficients (e.g., conductivity (k), conductance (C), etc.) cannot be added; however, their reciprocals, the resistances, can be added. When a R- or U-value falls between the numbers in the table, use the closest number to find the corresponding U- or R-value.

Resistance (or U-value)	U-value (or Resistance)	Resistance (or U-value)	U-value (or Resistance)
1.00	1.00	3.13	0.32
1.05	0.95	3.33	0.30
1.11	0.90	3.57	0.28
1.18	0.85	3.85	0.26
1.25	0.80	4.17	0.24
1.33	0.75	4.55	0.22
1.43	0.70	5.00	0.20
1.54	0.65	5.56	0.18
1.67	0.60	6.25	0.16
1.82	0.55	7.14	0.14
2.00	0.50	8.33	0.12
2.08	0.48	10.00	0.10
2.17	0.46	11.11	0.09
2.27	0.44	12.50	0.08
2.38	0.42	14.29	0.07
2.50	0.40	16.67	0.06
2.63	0.38	20.00	0.05
2.78	0.36	25.00	0.04
2.94	0.34	33.33	0.03

Example: Estimate the improvement in the U-value of a 4-in. concrete slab flat roof from adding a 4-in. glass-fiber batt (R-11) to the plenum space above the suspended plaster ceiling.

$$R$$

4-in. concrete slab roof with   7.14
$U = 0.14$

4-in. glass-fiber batt   + 11.11

Total R = 18.25 hrs/Btu/ft² /°F

Improved U-value will be 0.06 from above table (or use graph on page 64 to directly estimate improved U-value).

BUILDING MATERIALS—INSULATION R-VALUES FOR DWELLING UNITS

Insulation values to achieve FHA criteria (FHA Minimum Property Standards No. 2600) for dwelling units can be found in the table below. The suggested resistance (R) ratings in the table are given at various climate regions for specific applications. Manufacturers of mineral wool batts and blankets, for example, identify their products by R-values marked on the package.

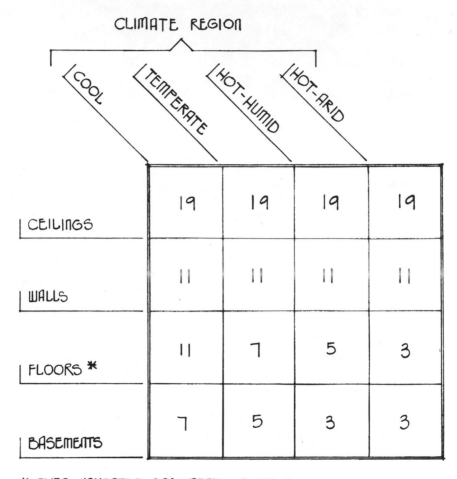

| | CLIMATE REGION | | | |
	COOL	TEMPERATE	HOT-HUMID	HOT-ARID
CEILINGS	19	19	19	19
WALLS	11	11	11	11
FLOORS *	11	7	5	3
BASEMENTS	7	5	3	3

* OVER UNHEATED BASEMENT OR UNHEATED CRAWL SPACES.

Note: FHA criteria are defined in terms of maximum *U*-values for exterior constructions. For typical residential frame wall and masonry wall constructions, however, the above table will give the minimum *R*-value needed from the insulation materials.

CHECKLIST FOR EFFECTIVE USE OF BUILDING MATERIALS

Insulation can reduce heat loss in winter and heat gain in summer so that smaller sized mechanical equipment will be required. This will conserve energy and reduce the overall cost of the building. Effective insulating materials are those which enclose, trap, or contain a film of air (e.g., glass-fiber, mineral wool, elastomeric foam). Insulating materials are commercially available in rigid boards, flexible blankets or batts, and even in loose fill. It is essential that thermal problems are recognized and solved at the early stages of a project.

Use moisture barriers to prevent condensation of moisture vapor within exterior constructions. Condensation within the building enclosure can destroy the insulating effectiveness of building materials. It also may freeze in winter months causing damage to the integrity of the structure.

Attic, crawl space, and wall ventilation will allow unwanted moisture to escape and help reduce heat buildup in summer months. In attic spaces, for example, the day-night temperature cycle can cause condensation on the underside of the roof when attics are not properly vented.

Prevent condensation on glass surfaces in cool and temperate regions by using double- or triple-glass or by locating a constant (not intermittent) warm-air supply near the floor under glass areas. The increased insulating effectiveness of double- and triple-glass will also lower building heating and cooling requirements.

BUILDING
HEAT
LOSS/GAIN

BUILDING HEAT LOSS/GAIN—HEAT LOSS IN WINTER

The major heat loss factors in winter are depicted below.

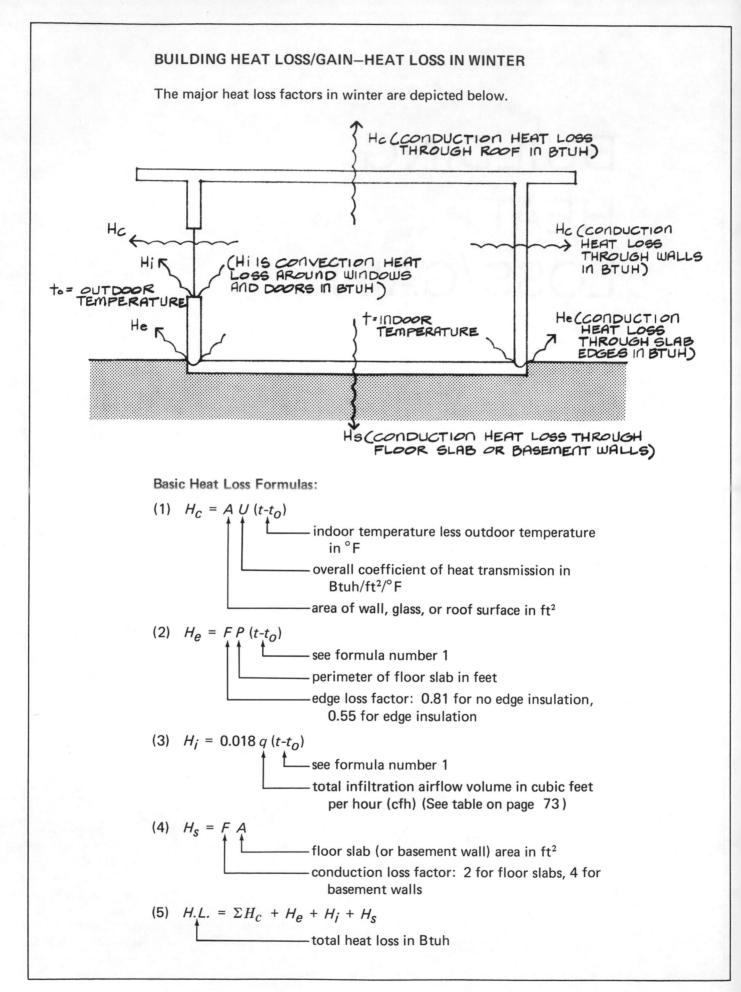

Basic Heat Loss Formulas:

(1) $H_c = A\,U\,(t-t_o)$

    indoor temperature less outdoor temperature in °F

    overall coefficient of heat transmission in Btuh/ft²/°F

    area of wall, glass, or roof surface in ft²

(2) $H_e = F\,P\,(t-t_o)$

    see formula number 1

    perimeter of floor slab in feet

    edge loss factor: 0.81 for no edge insulation, 0.55 for edge insulation

(3) $H_i = 0.018\,q\,(t-t_o)$

    see formula number 1

    total infiltration airflow volume in cubic feet per hour (cfh) (See table on page 73)

(4) $H_s = F\,A$

    floor slab (or basement wall) area in ft²

    conduction loss factor: 2 for floor slabs, 4 for basement walls

(5) $H.L. = \Sigma H_c + H_e + H_i + H_s$

    total heat loss in Btuh

RECOMMENDED OUTDOOR DESIGN TEMPERATURE AND DEGREE DAYS - HEATING

State	City	Temperature (°F)*	Degree Days**
AL	Birmingham	10	2,551
AK	Fairbanks	-60	14,279
AZ	Flagstaff	-5	7,152
	Phoenix	35	1,765
AR	Little Rock	10	3,219
CA	Los Angeles	40	2,061
	San Diego	45	1,458
	San Francisco	35	3,015
CO	Denver	-10	6,283
CT	Hartford	0	6,235
DE	Wilmington	5	4,930
DC	Washington	10	4,224
FL	Miami	45	214
	Tallahassee	25	1,485
GA	Atlanta	10	2,961
	Savannah	25	1,819
HI	Honolulu	60	0
ID	Boise	-10	5,809
IL	Chicago	-10	6,639
IN	Indianapolis	-10	5,699
IA	Des Moines	-15	6,588
KS	Wichita	-5	4,620
KY	Louisville	0	4,660
LA	Baton Rouge	20	1,560
	New Orleans	25	1,385
ME	Portland	-10	7,511
MD	Baltimore	10	4,654
MA	Boston	0	5,634
MI	Detroit	-5	6,293
MN	Minneapolis	-25	8,382
MS	Jackson	15	2,239
MO	St. Louis	-5	5,484
MT	Billings	-30	7,049
NB	Lincoln	-15	5,864
NV	Las Vegas	10	2,709
NH	Manchester	-10	7,383
NJ	Atlantic City	10	4,812
	Newark	0	4,589
	Trenton	0	4,980
NM	Albuquerque	10	4,348
NY	Albany	-10	6,875
	New York	5	5,219
NC	Greensboro	10	3,805
ND	Bismarck	-30	8,851
OH	Cincinnati	-5	4,410
OK	Tulsa	0	3,860
OR	Portland	10	4,635

RECOMMENDED OUTDOOR DESIGN TEMPERATURE AND
DEGREE DAYS HEATING (Continued)

State	City	Temperature ($°F$)*	Degree Days**
PA	Philadelphia	5	5,144
	Pittsburgh	-5	5,987
RI	Providence	0	5,954
SC	Columbia	20	2,484
	Greenville	10	2,980
SD	Rapid City	-20	7,345
TN	Chattanooga	10	3,254
	Knoxville	5	3,494
	Nashville	5	3,578
TX	Dallas	10	2,363
	Houston	20	1,396
UT	Salt Lake City	0	6,052
VT	Burlington	-15	8,269
VA	Charlottesville	10	4,166
	Richmond	10	3,865
WA	Seattle	15	4,424
WV	Charleston	0	4,476
WI	Milwaukee	-15	7,635
WY	Cheyenne	-20	7,381

*Environmental temperature data is also available from publications of the National Weather Service and the American Society of Heating, Refrigerating and Air-Conditioning Engineers (ASHRAE). Note that urban areas are always warmer than their surroundings due to heat storage capacity of masonry materials, heat loss from heated buildings, etc. In addition, there can be considerable variations in design temperatures for different locations within a state—especially where altitudes vary widely.

**The degree day is a unit that can be used to indicate the effect of climate on heating demand for buildings in winter. It is based on the temperature differential below $65°F$ in a 24 hour period (because heating demand for commercial buildings is usually zero at a mean, or average, daily temperature of $65°F$). For example, on a day when the average outdoor temperature is $45°F$, the degree days will be 65 - 45 = 20. The data given here represent typical yearly degree day totals as reported by airport weather stations.

AIR INFILTRATION

The table below lists infiltration airflow volume in cubic feet per hour (cfh) entering *per foot of crack* between sash and frame, or between door and frame, on the windward exposure. Since no more than two exterior walls of a room can face the wind, use the foot of crack value from the two adjacent walls having the most openings for total heat loss analysis. The infiltration values vary with the wind velocity as shown by the table.

Construction	Infiltration Airflow Volume (in cfh per ft of crack)		
	10 mph	20 mph	30 mph
Double-hung wood frame window:			
Non-weatherstripped	21	59	104
Weatherstripped	13	36	63
Double-hung metal frame window:			
Non-weatherstripped	47	104	170
Weatherstripped	19	46	76
Rolled steel frame glazed window or door (non-weatherstripped):			
Industrial, pivoted (or awning)	108	244	372
Residential, casement (or hinged)	18	47	74
Wood or metal door:			
Non-weatherstripped	69	154	249
Weatherstripped	19	51	92

DOUBLE-HUNG

PIVOTED

CASEMENT

SWINGING

Note: Reduce values above by $\frac{1}{3}$ (i.e., multiply by $\frac{2}{3}$) when storm window (or door) is added to weatherstripped construction, by $\frac{1}{2}$ when added to non-weatherstripped construction.

Infiltration through revolving and swinging door entrances depends on the pedestrian traffic rate, indoor-outdoor temperature difference, etc. The evaluation procedure requires experience; therefore, infiltration should be estimated only by qualified mechanical engineers. For preliminary estimates of main entrances in commercial buildings, however, use a swinging door infiltration rate of 600 cfh (at 10 mph wind conditions) times the number of door openings per hour.

BUILDING HEAT LOSS/GAIN—FLOOR SLABS

In winter, the heat loss to the ground through basement floor slabs and walls depends on the temperature of the adjacent earth (or ground water). Estimate this heat loss at about 2 Btuh/sq ft through floor slabs and 4 Btuh/sq ft for basement walls. The heat loss of the floor slab edge, however, is affected by the outdoor air temperature and can be greatly reduced by edge insulation and/or perimeter heating. Floor slab edge heat loss (H_e) can be found by the formula:

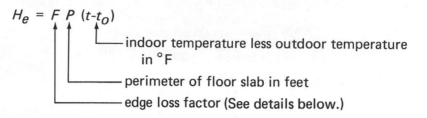

$$H_e = F \, P \, (t - t_o)$$

- indoor temperature less outdoor temperature in °F
- perimeter of floor slab in feet
- edge loss factor (See details below.)

Edge Insulation or Perimeter Heating ($F = 0.55$)

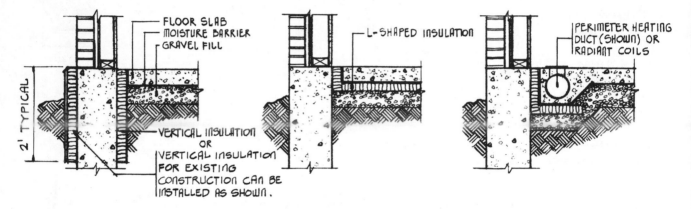

Note: To prevent cold floors always provide insulated slab edges. Typical width of 2 in. thick moisture resistant insulation is 2 ft below the outside grade line (or to 1 ft under the floor slab).

No Edge Insulation ($F = 0.81$)

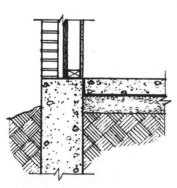
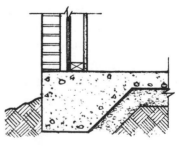

Note: For heat gain analyses, assume no heat gain through floor slab (and below-grade walls) in summer as ground temperature is generally equal to, or less than, room air temperature.

GUIDELINES FOR HEAT LOSS ANALYSIS

Establish Design Conditions

1. Outdoor winter design temperature (t_o) data is available from publications by the National Weather Service and the American Society of Heating, Refrigerating and Air-Conditioning Engineers (ASHRAE). The design day temperature is the winter temperature (December through March) that is equalled or exceeded about 98% of the time.

2. Indoor design temperature (t) is based on space use. For example, see table on page 10.

Identify Heat Loss Factors

1. Calculate, or find in tables, the *U*-values for: (a) exterior walls, (b) roof, (c) doors, and (d) glass. *U*-values give the heat flow in Btuh per sq ft when outdoor and indoor temperatures differ by 1°F. For floor constructions above grade, estimate the temperature difference between occupied rooms and ventilated crawl spaces at $\frac{2}{3}(t - t_o)$.

2. Estimate the heat loss through floor slabs, floor slab edges, and basement walls below grade.

3. Estimate the air infiltration or ventilation heat loss by the "crack method" which gives volume of air in cu ft per hr (cfh) entering per foot of crack based on the outdoor wind velocity. For fixed, sealed glass, however, assume an infiltration rate of about 15 cfh per square foot of glass surface area; for moveable sash windows about 30 cfh.

Heat Losses in Btuh

1. Calculate the heat loss through all roof and exterior opaque and glass wall constructions:

$$H_c = A\,U(t - t_o)$$

Note: Do *not* assume any beneficial heat gain from solar radiation through glass during winter months. During cloudless days, however, buildings with glazed openings will conserve heating energy when solar radiation can penetrate, warming inside air.

2. Calculate the floor slab edge heat loss:

$$H_e = F\,P\,(t - t_o)$$

Note: F is factor of 0.81 for floors without edge insulation; of 0.55 for floors with edge insulation.

3. Calculate the infiltration heat loss:

$$H_i = 0.018\,q\,(t - t_o)$$

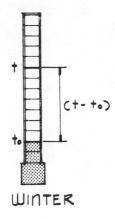

WINTER

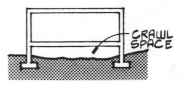

CRAWL SPACE

4. Calculate the heat loss through floor slab bottoms and basement walls:

$$H_s = F A$$

Note: F is factor of 2.0 for floor slabs; of 4.0 for basement walls.

Total Building Heat Loss in Btuh

Add the heat losses from the steps listed under the heading Heat Losses in Btuh. This figure is to be used for estimating the size of air-handling units (and heating elements as well). Heating Q is found by the formula:

$$Q = \frac{H.L.}{1.08 \, (t_e - t)}$$

where $H.L.$ = heat loss (room or building) in Btuh

t_e = temperature at heating equipment in °F

t = room or building temperature in °F

BUILDING HEAT LOSS/GAIN—EXAMPLE PROBLEM: HEAT LOSS

<u>GIVEN</u>:

BANK BRANCH OFFICE 30 FT × 15 FT × 10 FT HIGH LOCATED IN CINCINNATTI, OHIO. OUTDOOR WINTER DESIGN TEMPERATURE IS −5°F. INDOOR TEMPERATURE IS 70°F. U-VALUES ARE:

$$U_{WALLS} = 0.09$$
$$U_{ROOF} = 0.10$$
$$U_{GLASS} = 1.13$$

WHAT IS THE TOTAL HEAT LOSS (H.L.) IN BTUH FOR THIS BUILDING?

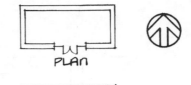

PLAN

ELEVATION

			AREA (FT²)	U	(t − t₀)	HEAT LOSS (BTUH)
① WALLS	n		30×10 = 300		(70−(−5)) = 75	
	E AND W		2×15×10 = 300			
	S (BRICK)		300 − 70 = <u>230</u>			
			830	0.09	75	5,603
	S (GLASS)		7 × 10 = 70	1.13	75	5,933
ROOF			30 × 15 = 450	0.10	75	3,375
						14,911

	PERIMETER (FT)	F	(t − t₀)	
② FLOOR SLAB EDGE: (2×30)+(2×15) = 90	0.55	75	3,713	
(WITH EDGE INSULATION) AREA (FT²)				
BOTTOM: 30×15 = 450	2		900	
	q	(t − t₀)		4,613
③ INFILTRATION	0.018	47×4×19	75	4,822
(DOORS AND WINDOWS)				4,822

<u>NOTE</u>: ASSUME A DESIGN WIND VELOCITY OF 20 MPH FOR ROLLED STEEL FRAME DOORS AND WINDOWS, NON-WEATHERSTRIPPED, HAVING 19 FT OF CRACK PER UNIT. HEAT LOSS THROUGH OPEN DOORS NOT INCLUDED. TO ACCOUNT FOR INFILTRATION FROM DOOR OPENINGS, USE 20/10 (600) = 1,200 CFH TIMES THE NUMBER OF DOOR OPENINGS PER HOUR (FROM TRAFFIC PREDICTIONS).

④ TOTAL HEAT LOSS   H.L. = $\boxed{24,346}$ BTUH

ASSUMING A TEMPERATURE (t$_e$) OF 130°F AT THE HEATING COILS, THE REQUIRED HEATING AIR FLOW VOLUME (Q) IN CUBIC FEET PER MINUTE (CFM) CAN BE FOUND AS FOLLOWS:

$$Q = \frac{H.L.}{1.08 (t_e - t)} = \frac{24,346}{1.08 (130 - 70)} = \boxed{375 \; CFM}$$

BE CAREFUL ALSO TO FIND THE Q FOR BUILDING VENTILATION AND FOR SUMMER COOLING AS THE HIGHEST Q VALUE FOR A SPACE WILL DETERMINE AIR-HANDLING UNIT AND DUCT SIZES.

BUILDING HEAT LOSS/GAIN—HEAT GAIN IN SUMMER

The major heat gain factors in summer are depicted below.

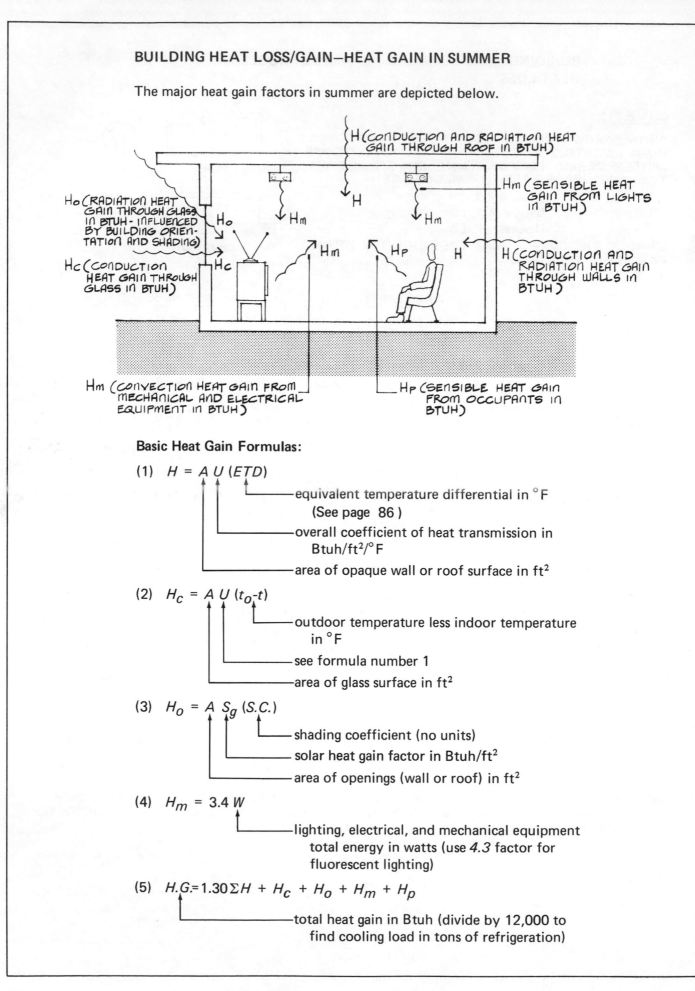

H (CONDUCTION AND RADIATION HEAT GAIN THROUGH ROOF IN BTUH)

Hm (SENSIBLE HEAT GAIN FROM LIGHTS IN BTUH)

Ho (RADIATION HEAT GAIN THROUGH GLASS IN BTUH - INFLUENCED BY BUILDING ORIENTATION AND SHADING)

Hc (CONDUCTION HEAT GAIN THROUGH GLASS IN BTUH)

H (CONDUCTION AND RADIATION HEAT GAIN THROUGH WALLS IN BTUH)

Hm (CONVECTION HEAT GAIN FROM MECHANICAL AND ELECTRICAL EQUIPMENT IN BTUH)

Hp (SENSIBLE HEAT GAIN FROM OCCUPANTS IN BTUH)

Basic Heat Gain Formulas:

(1) $H = A\,U\,(ETD)$

— equivalent temperature differential in °F (See page 86)

— overall coefficient of heat transmission in Btuh/ft²/°F

— area of opaque wall or roof surface in ft²

(2) $H_c = A\,U\,(t_o - t)$

— outdoor temperature less indoor temperature in °F

— see formula number 1

— area of glass surface in ft²

(3) $H_o = A\,S_g\,(S.C.)$

— shading coefficient (no units)

— solar heat gain factor in Btuh/ft²

— area of openings (wall or roof) in ft²

(4) $H_m = 3.4\,W$

— lighting, electrical, and mechanical equipment total energy in watts (use 4.3 factor for fluorescent lighting)

(5) $H.G. = 1.30\,\Sigma H + H_c + H_o + H_m + H_p$

— total heat gain in Btuh (divide by 12,000 to find cooling load in tons of refrigeration)

RECOMMENDED OUTDOOR DESIGN TEMPERATURE AND RELATIVE HUMIDITY (RH) - COOLING

State	City	Temperature (°F)	RH (%)
AL	Birmingham	95	48
AK	Fairbanks	78	43
AZ	Flagstaff	85	28
	Phoenix	105	27
AR	Little Rock	95	48
CA	Los Angeles	90	37
	San Diego	80	55
	San Francisco	80	45
CO	Denver	95	20
CT	Hartford	90	50
DE	Wilmington	90	59
DC	Washington	90	59
FL	Miami	90	62
	Tallahassee	95	48
GA	Atlanta	95	42
	Savannah	95	48
HI	Honolulu	85	60
ID	Boise	95	18
IL	Chicago	95	40
IN	Indianapolis	95	45
IA	Des Moines	95	48
KS	Wichita	100	32
KY	Louisville	95	48
LA	Baton Rouge	95	53
	New Orleans	95	53
ME	Portland	85	57
MD	Baltimore	90	59
MA	Boston	85	60
MI	Detroit	90	49
MN	Minneapolis	90	53
MS	Jackson	95	48
MO	St. Louis	95	48
MT	Billings	90	27
NB	Lincoln	95	48
NV	Las Vegas	110	14
NH	Manchester	85	60
NJ	Atlantic City	90	59
	Newark	90	53
	Trenton	90	59
NM	Albuquerque	95	22
NY	Albany	90	48
	New York	90	53
NC	Greensboro	90	53
ND	Bismarck	95	35
OH	Cincinnati	95	48
OK	Tulsa	100	36
OR	Portland	85	42

RECOMMENDED OUTDOOR DESIGN TEMPERATURE AND RELATIVE HUMIDITY (RH)—COOLING (Continued)

State	City	Temperature (°F)	RH (%)
PA	Philadelphia	90	59
	Pittsburgh	90	50
RI	Providence	90	50
SC	Columbia	95	48
	Greenville	95	40
SD	Rapid City	95	28
TN	Chattanooga	95	42
	Knoxville	95	40
	Nashville	95	48
TX	Dallas	100	38
	Houston	95	53
UT	Salt Lake City	95	18
VT	Burlington	90	45
VA	Charlottesville	90	59
	Richmond	90	59
WA	Seattle	80	45
WV	Charleston	90	50
WI	Milwaukee	90	50
WY	Cheyenne	90	22

Design values for summer outdoor conditions generally represent mid-afternoon averages. For other times, consult the environmental temperature and humidity data available from publications of the National Weather Service (e.g., National Climatic Center, Asheville, North Carolina 28801) or the American Society of Heating, Refrigerating and Air-Conditioning Engineers. Note that summer wet bulb temperature for equipment specification can be found from dry bulb temperature and R.H. given above by using the psychrometric chart.

HEAT GAIN FROM PEOPLE

Heat gain (Btuh per person at 75°F)

Typical Activity	Sensible*	Latent**	Total	Example Application
Seated at rest	245	155	400	grade schools, theaters, etc.
Office worker moderately active	250	200	450	offices, hotels, motels, apartments, college classrooms, etc.
Typewriting	250	250	500	general secretarial areas, stenographic pools, etc.
Clerk moderately active, standing at counter	275	275	550	banks, retail shops and stores, etc.
Dancing	305	545	850	dance halls, discotheques, etc.
Waiter serving in restaurant	375	625	1,000	cafeterias, restaurants, etc.
Bowling	580	870	1,450	bowling alleys, factories, industrial shops, etc.

*Sensible (or "dry") heat is the heat that can be measured by a change in temperature—"sensible" because it can be observed by the senses of sight and touch.

**Latent (or "wet") heat is the heat required to change the state of a substance. For example, it takes about 1 Btu to heat 1 lb of water from 211 to 212°F, but to change this 1 lb of water from a liquid at 212°F to steam at 212°F takes about 1,000 Btu. The 1,000 Btu is the latent heat in this example. In buildings, latent heat in the form of moisture changes the condition of the room air but not the temperature.

BUILDING HEAT LOSS/GAIN—HEAT GAIN FROM APPLIANCES

The cooling load for buildings must include the heat gain from all appliances—electric, gas, and steam. Shown below are some typical appliance sources of heat. If appliances are to be located in separate rooms (e.g., kitchens) that are *not* part of the air-conditioned area, the cooling load will be considerably reduced.

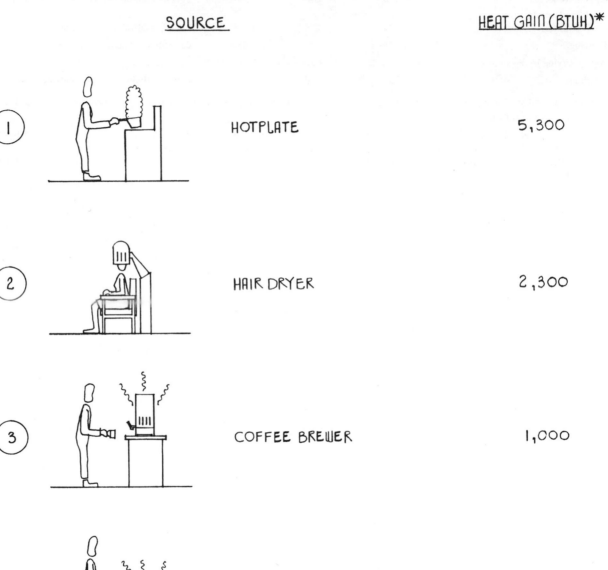

SOURCE	HEAT GAIN (BTUH)*
① HOTPLATE	5,300
② HAIR DRYER	2,300
③ COFFEE BREWER	1,000
④ INSTRUMENT STERILIZER	650

*BTUH'S ARE SENSIBLE HEAT GAIN FIGURES. SEE THE FOLLOWING PAGE FOR SENSIBLE AND LATENT HEAT GAIN BREAKDOWNS. FOR AN EXTENSIVE LIST OF APPLIANCE HEAT GAIN IN AIR-CONDITIONED AREAS (WITH AND WITHOUT EXHAUST HOODS), SEE ASHRAE HANDBOOK OF FUNDAMENTALS (1972), p. 417-19.

HEAT GAIN FROM APPLIANCES

Typical Appliance	Heat gain (Btuh)			Example Application
	Sensible	Latent	Total*	
Bunsen burner, $\frac{7}{16}$ in. barrel	1,680	420	2,100	laboratories, hospitals, etc.
Coffee brewer and warmer, electric, $\frac{1}{2}$ gal capacity, 625 watts	1,000	300	1,300	cafeterias, restaurants, residences, offices, etc.
Deep fat fryer, 14 lbs fat, 13 in. X 22 in. X 10 in. ht, 5500 watts	2,800	6,600	9,400	cafeterias, restaurants, etc.
Doughnut machine, 22 in. X 22 in. X 57 in. ht, 4700 watts	5,000	0	5,000	cafeterias, restaurants, etc.
Hairdryer, blower type, 1580 watts	2,300	400	2,700	beauty parlors, residences, etc.
Hotplate, 18 in. X 20 in. X 13 in. ht, 2 heating units, 5200 watts	5,300	3,600	8,900	cafeterias, restaurants, etc.
Roll warmer, 18 in. X 20 in. X 13 in. ht, 1650 watts	2,600	200	2,800	cafeterias, restaurants, etc.
Sterilizer, instrument, 1100 watts	650	1,200	1,850	hospitals, etc.
Toaster, four slice, pop-up, 2540 watts	2,230	1,970	4,200	cafeterias, restaurants, residences, etc.
Waffle iron, two grids, 1650 watts	1,680	1,120	2,800	cafeterias, restaurants, residences, etc.

*Heat gain can be reduced by 50% or more where effective exhaust hoods are provided.

U- AND TIME-LAG VALUES

Materials with low *U*-values are those which enclose, trap, or contain a film of air and generally are lightweight. On the other hand, materials with long thermal time-lags (i.e., building temperature lags behind outdoor temperature) are dense and heavyweight. Consequently, massive constructions will tend to produce more stable conditions (e.g., massive west walls, east walls, and roofs can greatly minimize solar heat impacts in summer); whereas, lightweight constructions are more sensitive to short-term solar impacts. Some data for homogeneous materials are given below.

Material	Thickness (in.)	U-value	Time-lag
Brick (common)	4	0.61	$2\frac{1}{2}$ hr
	8	0.41	$5\frac{1}{2}$ hr
	12	0.31	$8\frac{1}{2}$ hr
Concrete (sand and gravel aggregate)*	4	0.85	$2\frac{1}{2}$ hr
	8	0.67	5 hr
	12	0.55	8 hr
Insulating fiberboard	2	0.16	40 min
	4	0.09	3 hr
Wood (fir, yellow pine, etc.)	$\frac{1}{2}$	0.68	10 min
	1	0.47	25 min
	2	0.30	1 hr

*The thermal inertia of concrete floor slabs prevents radiant panel systems from effectively responding to rapid changes in room temperature demand.

The concept of equivalent temperature differential (*ETD*) is used to account for heat transmission (*U*-value) and thermal time-lag properties of materials.

EFFECT OF COLOR

Light colors tend to reduce building heat gain in summer. Accordingly, light colored walls of heavy mass will have the lowest *ETD* values (See following page). Most farmers have traditionally used white paint for their houses but used the less expensive red paint for their barns. The table below lists typical building finishes and ground coverings in order of increasing heat reflectivity.

Material	% Total Incident Heat Reflected
Tar & gravel, asphalt, etc.	7
Slate, dark soil, etc.	15
Grass, dry	30
Copper foil:	
tarnished	36
new	75
Paint:	
light gray	25
red	26
aluminum	46
light green	50
light cream	65
white	75
Whitewash, fresh snow cover, etc.	80
Aluminum foil*	95

*Can be used within exterior constructions to prevent radiant heat flow and to prevent moisture migration. Heat gain through ceilings can be further reduced by insulating materials having a reflective top surface.

Note: Summer air temperatures immediately above paved asphalt ground coverings can be 20°F or more higher than nearby shaded areas of grass.

EQUIVALENT TEMPERATURE DIFFERENTIAL (*ETD*) VALUES

Equivalent temperature differential (*ETD*) values in °F for opaque walls and roofs approximate the complex interrelationship between a construction's conductance, thermal time-lag, color, etc. The *ETD* is defined as the outdoor-indoor temperature difference that will be equal to the solar, convection, and radiation heat flow into a space (with allowance for time-lag). Shown below are typical values for opaque wall and roof constructions at the 40° north latitude. The data are based on a design room air temperature of 75°F.

EQUIVALENT TEMPERATURE DIFFERENTIALS *(ETD)* FOR WALLS AND ROOFS

	8 a.m. Dark	8 a.m. Light*	Noon Dark	Noon Light*	4 p.m. Dark	4 p.m. Light*	8 p.m. Dark	8 p.m. Light*
Lightweight construction (e.g., wood frame):								
N	15	8	19	13	27	21	27	21
NE	37	18	30	19	28	22	19	16
E	41	19	53	28	29	23	21	17
SE	24	12	55	31	30	24	20	17
S	7	3	37	22	45	30	19	16
SW	9	4	22	14	64	40	37	25
W	8	4	21	14	61	38	53	33
NW	7	4	20	14	39	27	49	31
Roof	12	2	77	39	86	48	25	16
Mediumweight construction (e.g., 4 in. brick or 4 in. concrete):								
N	8	4	17	12	27	21	25	20
NE	25	13	34	20	31	23	22	18
E	30	15	53	29	36	26	23	19
SE	19	9	52	29	39	27	24	19
S	3	1	32	19	47	31	26	20
SW	3	1	18	12	59	37	45	30
W	3	2	16	11	56	36	54	35
NW	3	1	16	11	40	27	45	30
Roof	1	0	59	28	88	48	44	27
Heavyweight construction (e.g., 8 in. brick or 8 in. concrete):								
N	9	7	10	7	15	11	21	16
NE	11	8	22	13	26	16	27	19
E	12	9	28	16	36	21	33	22
SE	11	8	22	13	35	20	34	22
S	10	8	11	7	25	16	32	21
SW	14	10	11	7	22	14	40	25
W	16	10	12	8	19	13	41	26
NW	13	9	11	7	16	11	32	21
Roof	15	7	25	11	46	23	50	27

*For cream colors, use values from light columns. For medium colors (e.g., medium blue, medium green, bright red, light brown, unpainted wood, natural concrete, etc.), use values halfway between light and dark. Dark blue, red, brown, green are considered dark colors.

Source: *ASHRAE Handbook of Fundamentals* (1967 and 1972 editions).

Note: Evaporating water can help cool roofs. Spraying the entire roof surface with water can reduce the heat gain through the roof by 20%. For a description of evaporative cooling techniques, see Steele, Alfred, "Roof Spraying," *Heating, Piping & Air Conditioning,* June 1968.

GUIDELINES FOR HEAT GAIN ANALYSIS

Establish Design Conditions

1. Outdoor summer design temperature (t_o) data is available from publications by the National Weather Service and the American Society of Heating, Refrigerating and Air-Conditioning Engineers (ASHRAE). The design day dry bulb temperature is the summer temperature (June through September) that is equalled or exceeded about 2% of the time.

2. Indoor design temperature (t) is based on space use.

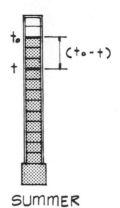

SUMMER

Identify Heat Gain Factors

1. Calculate, or find in tables, the U-values for: (a) exterior walls, (b) roof, (c) doors, and (d) glass.

2. Estimate the equivalent temperature differentials (ETD) for the opaque building constructions. The ETD concept is used to compensate for thermal time-lag as well as solar radiation effects.

3. Use data from tables for approximate solar heat gain factor (S_g) for clear, unshaded glass. Try to orient enclosure openings to minimize direct solar radiation or use shading devices if the openings must be on an unfavorable exposure. Shading coefficients $(S.C.)$ for interior blinds and shades, heat-absorbing glass, landscaping, and solar screens are available from tables. (For examples, see page 32.)

4. The heat gain from occupants depends on the density (i.e., number of occupants) and type of activity. The table on page 81 can be used to establish sensible heat gains in Btuh per person.

Note: Air-conditioning systems increase the air pressure in buildings by adding treated outdoor air to the conditioned spaces. This prevents air infiltration except at building entrances. For preliminary estimates, use a sensible heat infiltration at entrances of 300 Btuh times the number of door openings per hour.

Heat Gains in Btuh (Environmental and Internal)

1. Calculate the heat gain (conduction and radiation) through all opaque roof and wall constructions:

$$H = A \, U(ETD)$$

2. Calculate the heat gains (conduction, H_c and radiation, H_o) through all openings:

(a) $$H_c = A \, U \, (t_o - t)$$

(b) $$H_o = A \, S_g \, (S.C.)$$

3. Calculate the internal heat gains from: (a) occupants, and (b) electrical-mechanical equipment:

(a) $$H_p = nM$$

where n = number of occupants

M = activity sensible heat gain in Btuh

(b) $$H_m = 3.4W$$

Light sources are marked with their watts (W). For preliminary estimates of heat gain from lighting systems, use 2 to 6 watts/sq ft. In office buildings, lighting typically produces about one-third of the total heat gain. If wattage is unavailable for other equipment, multiply amperage by the voltage for a rough estimate.

Total Building Heat Gain in Btuh

Add the heat gains from the steps listed under the heading Heat Gains in Btuh (Environmental and Internal). This figure is to be used for estimating the size of air-handling units and ducts since $Q \simeq H$ (sensible)/22 for cooling. As a preliminary estimate accounting for sensible and latent heat loads, increase the sensible total by 30% (i.e., multiply total by 1.30 factor; by 1.10 in hot-arid region). This figure can be used to estimate the size of the refrigeration elements.

Note: The cooling capacity of refrigeration equipment is often expressed in tons. This ton unit does *not* mean the weight of the equipment. It originated with the cooling systems that blew air over melting ice. The heat required to change 1 lb of ice into 1 lb of water at 32°F is about 144 Btu. The melting of a ton of ice in 24 hours, therefore, will cool (2,000 lbs X 144 Btu)/24 hrs. = 12,000 Btuh.

BUILDING HEAT LOSS/GAIN—EXAMPLE PROBLEM: HEAT GAIN

GIVEN:

RETAIL STORE 40FT × 20FT × 10FT HIGH LOCATED IN PHILADELPHIA, PA. OUTDOOR SUMMER DESIGN TEMPERATURE IS 90°F. INDOOR IS 75°F. U-VALUES ARE:

U WALLS = 0.07 (HEAVYWEIGHT, DARK COLOR CONSTRUCTION)

U ROOF = 0.08 (MEDIUMWEIGHT, LIGHT COLOR CONSTRUCTION)

U GLASS = 1.06

PLAN

ELEVATION

WHAT IS THE TOTAL HEAT GAIN (H.G.) IN BTUH FOR THIS BUILDING?

		MATERIAL		AREA (FT²)	U	ETD (SEE PAGE 86)	HEAT GAIN (BTUH)	
①	OPAQUE WALLS	N	BRICK	40×10 = 400	0.07	15	420	
		E	BRICK	20×10 = 200	0.07	36	504	
		S	GLASS	SEE STEP 2 BELOW				
		W	GLASS	SEE STEP 2 BELOW				
			BRICK	100	0.07	19	133	
	ROOF		CONCRETE	40×20 = 800	0.08	48	3,072	

NOTE: NO HEAT GAIN THROUGH FLOOR SLAB AS GROUND TEMPERATURE IN SUMMER IS GENERALLY EQUAL TO, OR LESS THAN, ROOM AIR TEMPERATURE.   4,129

			AREA (FT²)	U	(tₒ-t)		
②	GLASS (CONDUCTION)	S	400		(90-75)		
		W	100				
			500	1.06	15	7,950	

7,950

			AREA (FT²)	Sg	S.C.		
③	GLASS (RADIATION)	S (WITH CONTINUOUS OVERHANG)	400	26	0.25	2,600	
		W (WITH CURTAINS)	100	194	0.60	11,640	

NOTE: CONSULT SOLAR TABLES TO SELECT THE TIME OF DAY ON 21 JUNE AT WHICH THE SOLAR HEAT GAIN THROUGH OPENINGS WILL BE GREATEST (I.E., 4PM FOR THIS EXAMPLE SHOWN ON PAGE 35).   14,240

④ LIGHTING: Hm = 4.3W = 4.3 (2,400 WATTS) =   10,320

PEOPLE: ASSUMING 10 OCCUPANTS: 10 × 250 = 2,500

12,820

⑤ TOTAL HEAT GAIN (SENSIBLE):   39,139

TOTAL HEAT GAIN (SENSIBLE + LATENT):   H.G. = 1.30 × 39,139 =   50,881 BTUH

OR ≃ 4 TONS

NOTE: USE PSYCHROMETRIC CHART TO FIND SENSIBLE AND LATENT HEAT GAINS, OR INCREASE SENSIBLE TOTAL BY 30% AS SHOWN ABOVE.

EQUIPMENT SELECTION BASED ON STEP 5 TOTAL HEAT GAIN (SENSIBLE + LATENT) ALONE, HOWEVER, GENERALLY WILL BE OVERSIZED. THIS IS DUE TO THE ASSUMPTION THAT THE MAXIMUM SOLAR, LIGHTING, AND PEOPLE HEAT GAINS OCCURRED AT ONE TIME. TO FIND A REALISTIC COOLING LOAD, REFER TO THE REDUCTION FACTORS IN CARRIER SYSTEM DESIGN MANUAL, PART NUMBER 1 ON "LOAD ESTIMATING".

BUILDING HEAT GAIN AND ZONING

Small buildings that have fairly steady occupancy (e.g., offices, residences), or buildings with a considerable amount of glass, will have a peak heat gain largely controlled by solar radiation. However, buildings that have variable occupancy (e.g., restaurants, theaters), generally will have a maximum total heat gain at the time of the greatest number of occupants. The design of the mechanical system should be based on the hour of the day at which the heat gain from the various sources—sun, lighting, people, etc.—is the greatest.

In large, multi-use buildings, there may be many different peak heat gains throughout the building. For example, a tall building may have large conference rooms with glass on the north exposure, interior office and secretarial spaces, and a cafeteria on the top floor with glass on the west exposure. Consequently, these buildings are often separated into areas or "zones" to handle the various cooling and heating requirements. Heat gain analysis can be rather complex requiring computation of the maximum heat gain for each zone.

BUILDING HEAT LOSS/GAIN—BUILDING HEAT GAIN EXAMPLES

Typical building heat gain load patterns, i.e., Btuh versus time of day, are shown below. Mechanical systems should be sized for cooling loads below the peak heat gains. Smaller equipment will have lower installation cost and, by operating for extended periods of time at peak capacity, are often the most economical selection. Heat gains for spaces along the building exterior are highly variable and usually greater than interior heat gains from lights, machines, and people.

Exterior Offices (Except North Side)

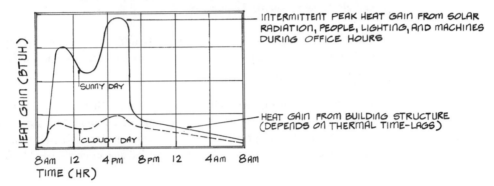

Retail Shops and Stores, Interior Offices, Factories

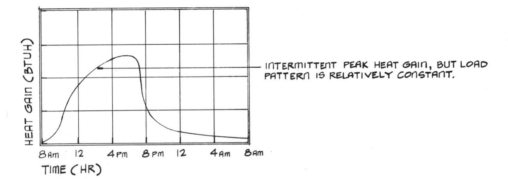

Apartment Buildings, Hotels, Hospitals, Residences

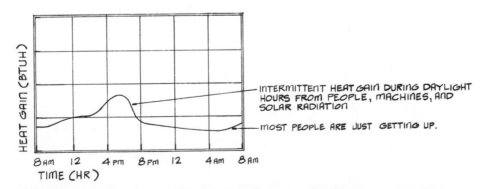

Note: The heat gain for residences is largely controlled by outdoor conditions. Cooling units, therefore, should be slightly undersized so that the indoor temperature will rise 3 to 5°F (called "swing" temperature) following the maximum outdoor temperature on the few design days, i.e., days when all loads peak simultaneously. Under normal conditions, however, the indoor temperature will be constant.

CHECKLIST FOR BUILDING HEAT LOSS/GAIN

For effective control of air infiltration, provisions should be made to seal cracks or openings in exterior constructions with caulking and weatherstripping. In public buildings, use revolving doors, vestibules, automatic door closures, etc. to reduce air infiltration at entrances. Try to orient entrances away from prevailing winter winds.

Floor slab edge heat loss can be reduced about one-third by edge insulation and perimeter heating.

When computing cooling loads, always include the heat gain from building occupants and electrical-mechanical equipment.

Use white or light-colored exterior walls and roofs to effectively reflect solar radiation. In cold climate regions, use dark-colored exterior walls and roofs to absorb solar radiation.

Separate buildings into zones based on specific heating and cooling demand requirements. To conserve energy, avoid heating (or cooling) corridors, lobbies, storage areas, etc. at the temperature level of occupied spaces. In addition, when located along the building perimeter, these unoccupied spaces can act as "buffer" zones.

MECHANICAL SYSTEMS

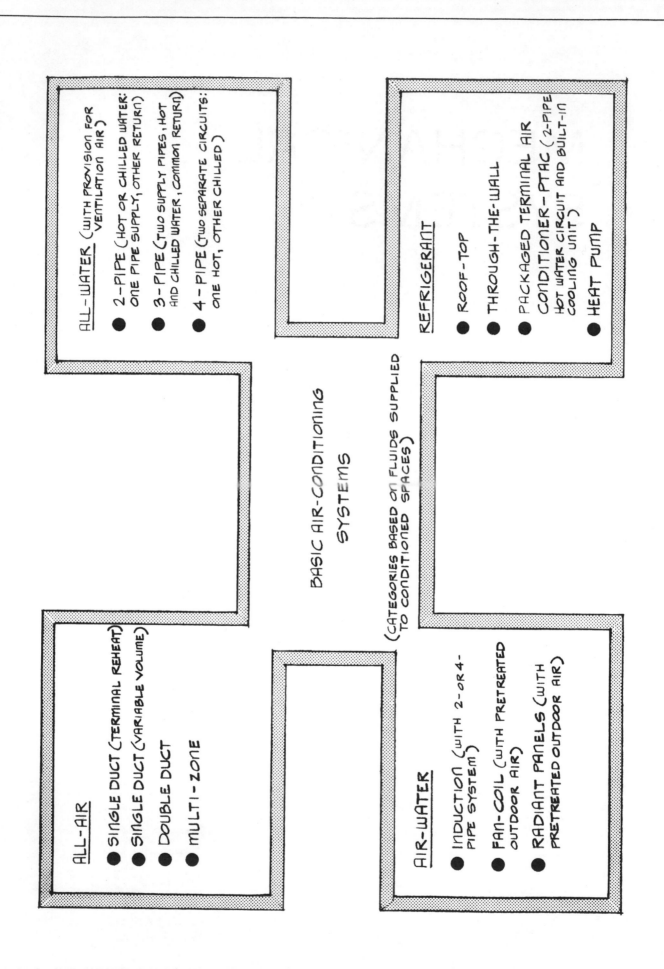

BASIC AIR-CONDITIONING SYSTEMS

(CATEGORIES BASED ON FLUIDS SUPPLIED TO CONDITIONED SPACES)

ALL-AIR

● SINGLE DUCT (TERMINAL REHEAT)
● SINGLE DUCT (VARIABLE VOLUME)
● DOUBLE DUCT
● MULTI-ZONE

ALL-WATER (WITH PROVISION FOR VENTILATION AIR)

● 2-PIPE (HOT OR CHILLED WATER: ONE PIPE SUPPLY, OTHER RETURN)
● 3-PIPE (TWO SUPPLY PIPES, HOT AND CHILLED WATER, COMMON RETURN)
● 4-PIPE (TWO SEPARATE CIRCUITS: ONE HOT, OTHER CHILLED)

AIR-WATER

● INDUCTION (WITH 2-OR 4-PIPE SYSTEM)
● FAN-COIL (WITH PRETREATED OUTDOOR AIR)
● RADIANT PANELS (WITH PRETREATED OUTDOOR AIR)

REFRIGERANT

● ROOF-TOP
● THROUGH-THE-WALL
● PACKAGED TERMINAL AIR CONDITIONER-PTAC (2-PIPE HOT WATER CIRCUIT AND BUILT-IN COOLING UNIT)
● HEAT PUMP

MECHANICAL SYSTEMS—HEATING AND COOLING

Example forced air heating and cooling units, suitable for small buildings, are described below.

Heating

Forced warm-air systems, as shown here, can deliver filtered, conditioned air to the building locations of greatest heat loss and in an air supply direction that will not disturb room occupants. Other heating methods employ fin-coil convectors (See page on hydronic systems) and radiant panels built into room surfaces.

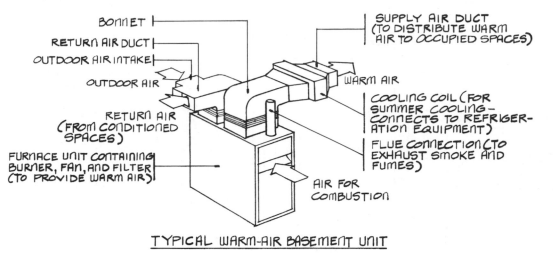

TYPICAL WARM-AIR BASEMENT UNIT

Note: In air-water and all-water systems, boilers provide hot water or steam. Steam is used for heating and for special process requirements (e.g., laundries, kitchens, sterilizers).

Cooling

The cooling system shown below has a fan-coil unit located above a suspended ceiling. The coil is cooled by chilled water from the "air-cooled chiller" unit located outdoors. The fan blows air over the coils to supply cold air to the occupied spaces.

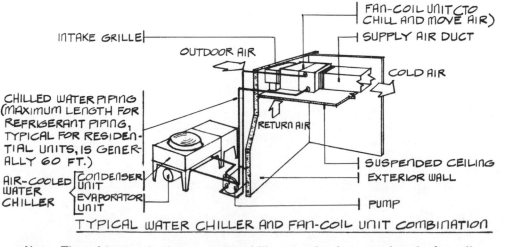

TYPICAL WATER CHILLER AND FAN-COIL UNIT COMBINATION

Note: The refrigerant in the evaporator chills water that is pumped to the fan-coil unit.

MECHANICAL SYSTEMS—HEATING: COMBUSTION PROCESS

In the combustion process, oil or gas is burned in a combustion chamber to provide heat. The air (or water for a boiler) supply is heated by conduction at the outside of the heat exchanger. A flue or chimney is required to exhaust the smoke and fumes of combustion to the atmosphere. Where forced draft fan boilers are used, be certain flue height extends above building roof line. The basic elements of a warm-air furnace are shown by the diagram below.

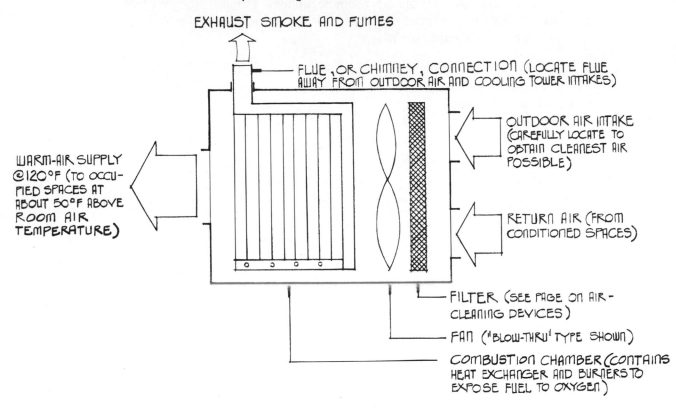

Note: Electric resistance strip heaters are available that work like large toasters. They replace the furnace section as the source of heat for the supply air.

MECHANICAL SYSTEMS—COOLING: COMPRESSIVE REFRIGERATION CYCLE

The process of removing heat from interior spaces to the outdoors is based on the principle of thermodynamics that heat will move to a cooler medium. Cooling by mechanical compression is shown by the diagram below. Heat is transferred from the chilled water system (See evaporator-cooling coil cycle) to the condenser water system (See condenser-cooling tower cycle) by means of changing the state of a refrigerant (See Freon compressor cycle) which gives off and absorbs heat.

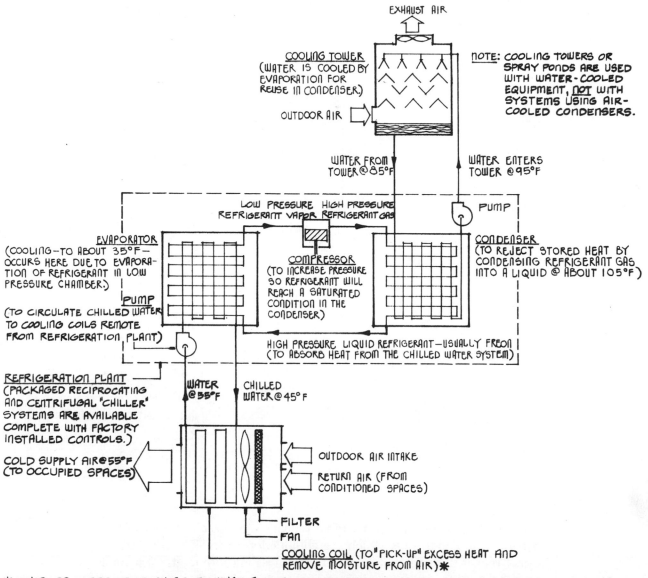

✳ THIS PROCESS IS SIMILAR TO WHAT NORMALLY OCCURS IN NATURE. FOR EXAMPLE, WHEN A WARM-AIR MASS MEETS A COLD FRONT THE AIR IS COOLED AND LOSES ITS CAPACITY TO HOLD MOISTURE. AS A CONSEQUENCE, THE MOISTURE CONDENSES AND FALLS AS RAIN.

MECHANICAL SYSTEMS—ALL-AIR: SINGLE DUCT SYSTEM

The single duct system supplies a single stream of either hot or cold conditioned air at low velocity. Its capacity to handle varying building loads can be adjusted by airflow volume controls called "dampers." A damper is an obstruction that can be used to vary the airflow in ducts. In a single duct system each branch should have its damper adjusted to supply the proper airflow volume—this process is called "balancing" the system. (See page on airflow control devices.)

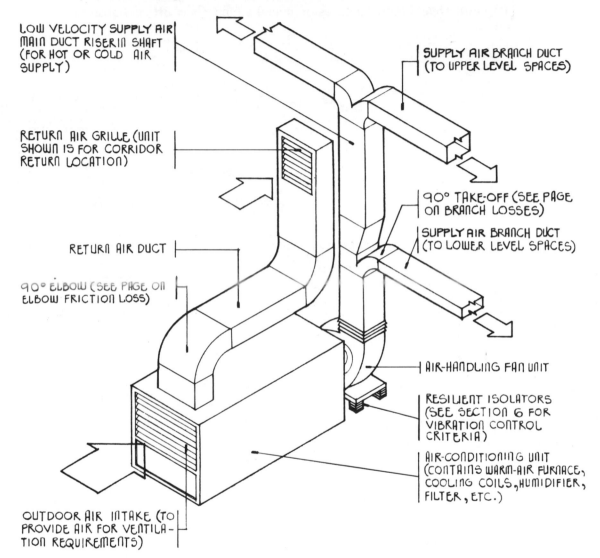

LOW VELOCITY SUPPLY AIR MAIN DUCT RISER IN SHAFT (FOR HOT OR COLD AIR SUPPLY)

RETURN AIR GRILLE (UNIT SHOWN IS FOR CORRIDOR RETURN LOCATION)

RETURN AIR DUCT

90° ELBOW (SEE PAGE ON ELBOW FRICTION LOSS)

OUTDOOR AIR INTAKE (TO PROVIDE AIR FOR VENTILATION REQUIREMENTS)

SUPPLY AIR BRANCH DUCT (TO UPPER LEVEL SPACES)

90° TAKE-OFF (SEE PAGE ON BRANCH LOSSES)

SUPPLY AIR BRANCH DUCT (TO LOWER LEVEL SPACES)

AIR-HANDLING FAN UNIT

RESILIENT ISOLATORS (SEE SECTION 6 FOR VIBRATION CONTROL CRITERIA)

AIR-CONDITIONING UNIT (CONTAINS WARM-AIR FURNACE, COOLING COILS, HUMIDIFIER, FILTER, ETC.)

A single duct system with "terminal reheat" supplies air at about 55°F to room (or zone) units which are equipped with small steam (or hot water) reheat coils to compensate for changes in room loads. Terminal reheat units can also use the rejected heat from refrigeration compressors.

Note: The variable air volume system is an all-air system that supplies a variable quantity of air at constant temperature to occupied spaces based on specific room requirements. Terminal control devices regulate airflow volume to match the variation in room load (e.g., by means of a bypass damper). Simultaneous cooling and heating, however, is not possible with the variable air volume system, but at partial load conditions it is more efficient than the reheat system.

MECHANICAL SYSTEMS—ALL-AIR: BASIC ELEMENTS OF A SINGLE DUCT SYSTEM

Shown below are the basic elements of the all-air single duct system. The arrows on this schematic diagram indicate the interrelationship of the system components.

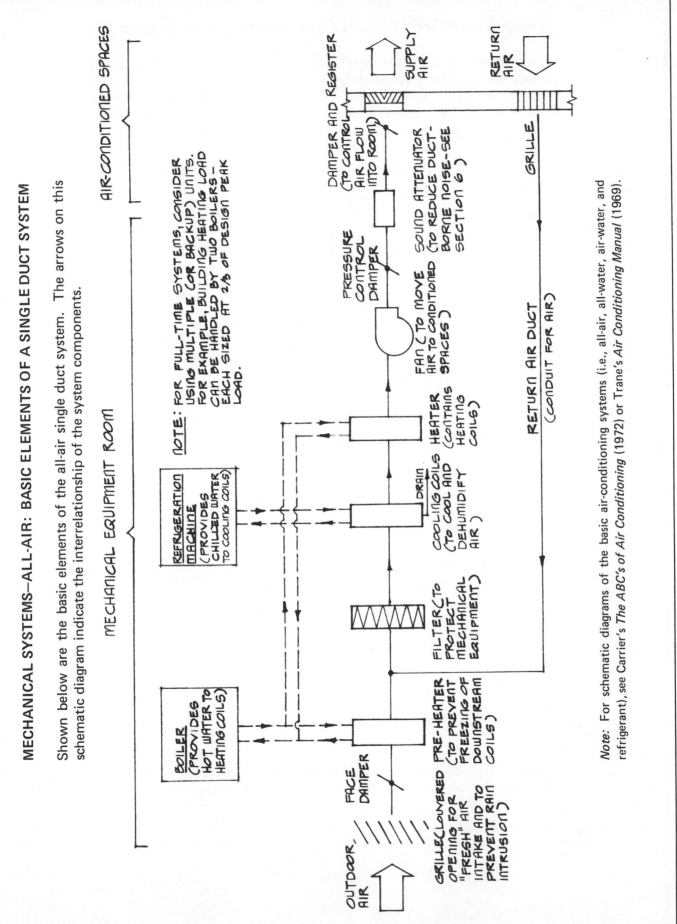

Note: For schematic diagrams of the basic air-conditioning systems (i.e., all-air, all-water, air-water, and refrigerant), see Carrier's *The ABC's of Air Conditioning* (1972) or Trane's *Air Conditioning Manual* (1969).

MECHANICAL SYSTEMS—ALL-AIR: DOUBLE DUCT SYSTEM

In the double (or dual) duct system, hot and cold air are supplied by separate ducts to a terminal called a "mixing box." A mixing box can serve one room, as shown below, or a zone consisting of several rooms. The mixing box usually has thermostatically controlled, adjustable dampers. It also should be designed to prevent the transmission of noise from air turbulence. To reduce ceiling and shaft space requirements, air speeds in double duct systems usually are in the range of 2,500 to 6,000 fpm.

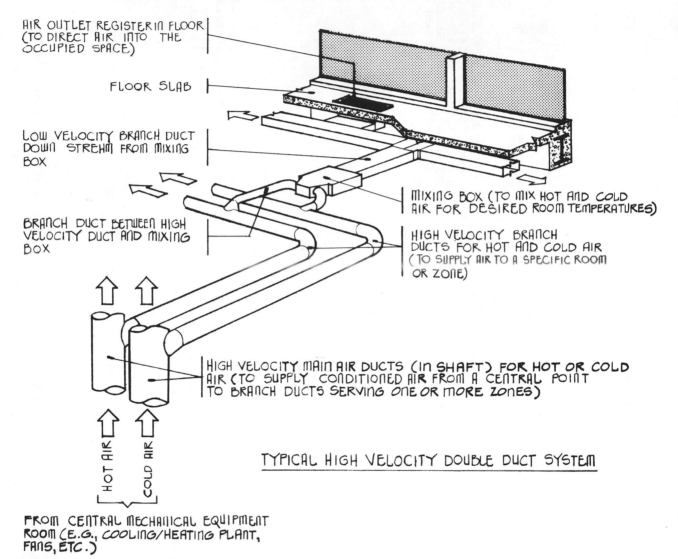

AIR OUTLET REGISTER IN FLOOR (TO DIRECT AIR INTO THE OCCUPIED SPACE)

FLOOR SLAB

LOW VELOCITY BRANCH DUCT DOWN STREAM FROM MIXING BOX

BRANCH DUCT BETWEEN HIGH VELOCITY DUCT AND MIXING BOX

MIXING BOX (TO MIX HOT AND COLD AIR FOR DESIRED ROOM TEMPERATURES)

HIGH VELOCITY BRANCH DUCTS FOR HOT AND COLD AIR (TO SUPPLY AIR TO A SPECIFIC ROOM OR ZONE)

HIGH VELOCITY MAIN AIR DUCTS (IN SHAFT) FOR HOT OR COLD AIR (TO SUPPLY CONDITIONED AIR FROM A CENTRAL POINT TO BRANCH DUCTS SERVING ONE OR MORE ZONES)

HOT AIR

COLD AIR

FROM CENTRAL MECHANICAL EQUIPMENT ROOM (E.G., COOLING/HEATING PLANT, FANS, ETC.)

TYPICAL HIGH VELOCITY DOUBLE DUCT SYSTEM

Note: Floor-ceiling systems (e.g., ribbed or corrugated structural decks) are also available that can be used to distribute air to registers or induction units.

MECHANICAL SYSTEMS—ALL-WATER: HYDRONIC SYSTEMS

The all-water (or hydronic) system heats and cools with water. Room ventilation, however, must be provided by a direct connection to the outdoors or a separate air duct system. The 2-pipe and 4-pipe systems are shown below. The 3-pipe system (hot and chilled water supply circuit and a common return) is inefficient where one side of a building needs heating and the other cooling. Note that hydronic systems will require less space than air systems as water is more efficient and compact than air as a medium for heat transfer.

2-Pipe System (Fan-Coil)

In the 2-pipe system, either hot *or* chilled water is supplied through one pipe to fin-coil or fan-coil units and one pipe returns water to the mechanical equipment.

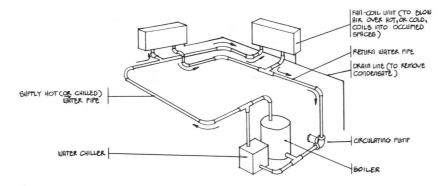

Note: Pipes should be encased in thermal insulation having an outside moisture barrier to prevent condensation.

4-Pipe System (Fan-Coil)

In the 4-pipe system, two supply circuits are used—one for hot water and the other for chilled water. This system has the flexibility to handle simultaneous demands for heating and cooling from different rooms or zones within a building. However, energy will be wasted when both hot and chilled water must be supplied to fan-coil units— previously chilled water is heated and previously heated water is cooled!

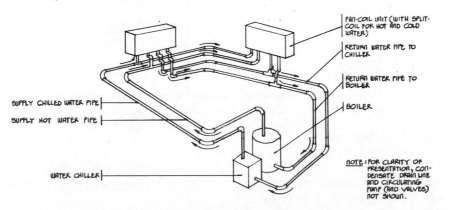

Note: Conventional identification of piping systems by color is: red—fire protection; yellow or orange—dangerous materials; and green (or white, black, gray)— safe materials.

MECHANICAL SYSTEMS—UNDER-GLASS CONVECTORS

The graph below may be used to help determine if heating convectors will be needed under exterior glazing to counteract downdrafts. For example, an office building having 7 ft high (*H*) double-glass curtain walls should have under-glass convectors if workers will be seated adjacent to the glass and if the outdoor winter design temperature is 18°F or below. (See dashed lines on graph.)

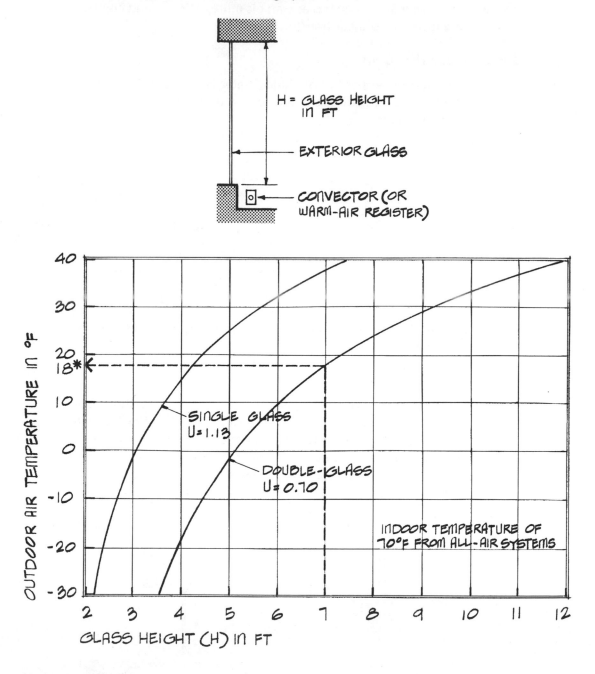

* DOWNDRAFTS NOT SIGNIFICANT FOR OUTDOOR TEMPERATURES ABOVE 18°F IN THIS DOUBLE-GLASS EXAMPLE.

MECHANICAL SYSTEMS—FIN-COILS

Typical fin-coil and fin-tube heating data are given below for units having an enclosure height (*H*) of 12 in. To find required fin-coil length for a space, divide room heat loss by the heat output per linear foot at the system's design water temperature. Consult manufacturer's catalogs, or Institute of Boiler and Radiator Manufacturers (I = B = R) publications, for specific output data from various enclosure heights, fin sizes, etc. Fin-coil heating can blanket the entire exterior wall surface preventing drafts. For example, underfloor units are often used where fixed glass (or sliding glass doors) extends to the floor.

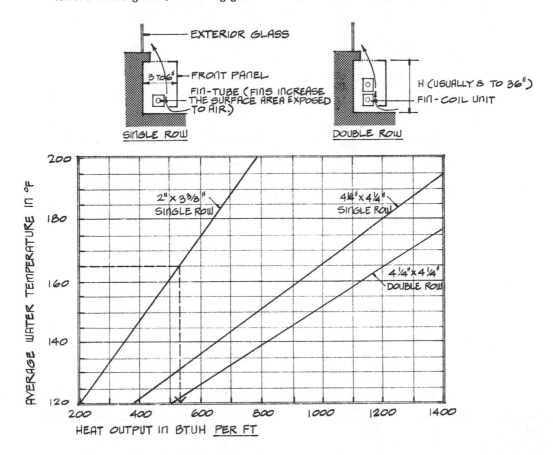

Example—Use of Graph

Given: Hydronic system for room having heat loss of 16,000 Btuh. Assume average water temperature of 165°F. Use single row 2 in. X 3 $\frac{3}{8}$ in. fin-tube with H = 12 in.

Procedure to size fin-tube:

1. Enter graph at 165°F and read opposite 2 in. X 3 $\frac{3}{8}$ in. curve to heat output of 525 Btuh per foot (See dashed lines).
2. Find required length, i.e., 16,000/525 ≃ 31 ft.

Note: Provide expansion devices in long pipe runs to allow movement and to prevent noise. Be sure also to carefully seal all pipe penetrations through floors, partitions, etc.

MECHANICAL SYSTEMS—AIR-WATER: INDUCTION SYSTEM

The induction system supplies air to the occupied space (usually perimeter zone) from a high velocity duct by induction nozzles. The air from the nozzles induces room airflow over a fin-coil unit. Depending on the time of year, chilled or hot water is pumped through the same pipes from a central point to the fin-coil. The two-pipe circuit, as shown below, therefore, provides only heating *or* cooling. The four-pipe and three-pipe systems, however, circulate hot and chilled water at all times for year-round air-conditioning. Note that perimeter zone systems must be able to respond to highly variable heating and cooling loads.

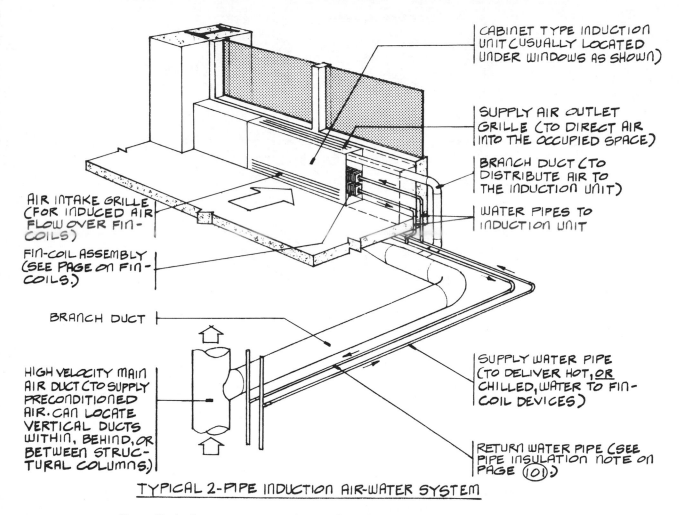

CABINET TYPE INDUCTION UNIT (USUALLY LOCATED UNDER WINDOWS AS SHOWN)

SUPPLY AIR OUTLET GRILLE (TO DIRECT AIR INTO THE OCCUPIED SPACE)

BRANCH DUCT (TO DISTRIBUTE AIR TO THE INDUCTION UNIT)

WATER PIPES TO INDUCTION UNIT

AIR INTAKE GRILLE (FOR INDUCED AIR FLOW OVER FIN-COILS)

FIN-COIL ASSEMBLY (SEE PAGE ON FIN-COILS)

BRANCH DUCT

HIGH VELOCITY MAIN AIR DUCT (TO SUPPLY PRECONDITIONED AIR. CAN LOCATE VERTICAL DUCTS WITHIN, BEHIND, OR BETWEEN STRUCTURAL COLUMNS.)

SUPPLY WATER PIPE (TO DELIVER HOT, OR CHILLED, WATER TO FIN-COIL DEVICES)

RETURN WATER PIPE (SEE PIPE INSULATION NOTE ON PAGE (101).)

TYPICAL 2-PIPE INDUCTION AIR-WATER SYSTEM

Note: Drain line to remove condensate from fin-coil not shown for clarity of presentation.

The 2-pipe fan-coil system is similar to the induction system shown above. Air is blown over the fin-coils by a fan (hence the term "fan-coil") so there is no single duct circuit used for air supply. Air for ventilation is often supplied by a direct through-the-wall connection.

MECHANICAL SYSTEMS—RADIANT PANEL HEATING

In radiant panel systems, room heat is furnished by heated floor, ceiling, or sometimes wall surfaces. The panel heat is supplied from air ducts (or narrow plenum areas), water piping, or electric resistance elements. Radiant panel systems can provide a high level of thermal comfort at low air temperatures, but can *not* respond quickly to changes in temperature demands. For example, radiant floor slabs have a thermal inertia that prevents rapid response when the outdoor temperature increases suddenly or when several people enter a room.

Ceiling Panel (Electric Resistance System)

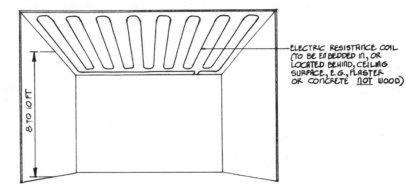

ELECTRIC RESISTANCE COIL (TO BE EMBEDDED IN, OR LOCATED BEHIND, CEILING SURFACE, E.G., PLASTER OR CONCRETE — NOT WOOD)

8 TO 10 FT

Note: The radiant panel system can not counteract downdrafts at cold glass surface areas. In addition, desk and table tops can interrupt radiant heat energy causing cold spots on the side opposite the panel source.

Typical Ceiling and Floor Details (Hot Water Systems Shown)

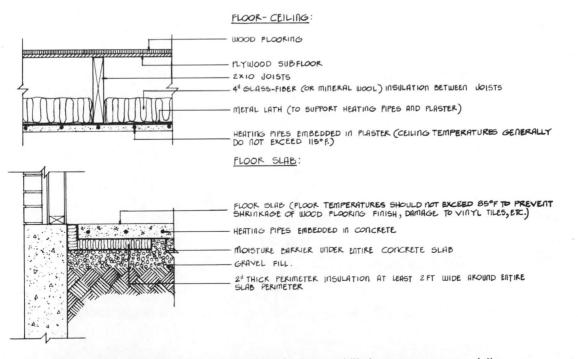

FLOOR-CEILING:

WOOD FLOORING

PLYWOOD SUBFLOOR

2x10 JOISTS

4" GLASS-FIBER (OR MINERAL WOOL) INSULATION BETWEEN JOISTS

METAL LATH (TO SUPPORT HEATING PIPES AND PLASTER)

HEATING PIPES EMBEDDED IN PLASTER (CEILING TEMPERATURES GENERALLY DO NOT EXCEED 115°F.)

FLOOR SLAB:

FLOOR SLAB (FLOOR TEMPERATURES SHOULD NOT EXCEED 85°F TO PREVENT SHRINKAGE OF WOOD FLOORING FINISH, DAMAGE TO VINYL TILES, ETC.)

HEATING PIPES EMBEDDED IN CONCRETE

MOISTURE BARRIER UNDER ENTIRE CONCRETE SLAB

GRAVEL FILL.

2" THICK PERIMETER INSULATION AT LEAST 2 FT WIDE AROUND ENTIRE SLAB PERIMETER

Note: Ceiling panels with integral tubing for hot or chilled water are commercially available for use in suspended lay-in systems. Supplementary air also is provided for ventilation and humidity control and to remove the heat from lights.

MECHANICAL SYSTEMS—REFRIGERANT: ROOF-TOP AND THROUGH-THE-WALL SYSTEMS

Refrigerant systems use self-contained units for extracting or adding heat.

Roof-Top Unit

The single zone roof-top unit shown below has an air-cooled condenser for cooling and a gas-fired furnace for heating. Low-rise buildings can be zoned for heating and cooling with roof-top units. Each unit can be sized to serve a specific zone, e.g., room, retail store, or even separate floor levels. Multi-zone and double duct roof-top units are also commercially available.

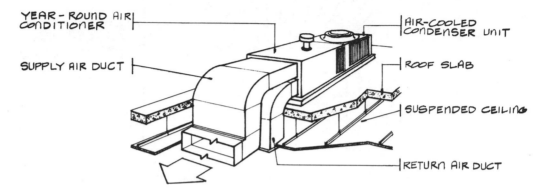

Note: Provide easy access to roof and work area adjacent to units for maintenance and equipment service. Units usually take up about 5 sq ft of roof area per ton of refrigeration.

Through-The-Wall Unit

The unit shown below provides cooling through a supply duct. Heating can be provided either by reverse cycle (i.e., heat pump) or by supplementary heating (e.g., electric resistance elements). Units are also available for use without ductwork to provide cooling and heating directly within the conditioned space.

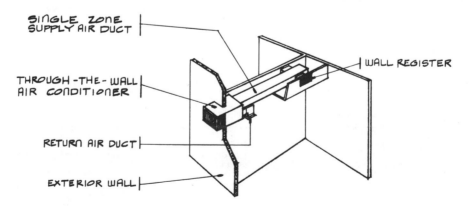

REFERENCE

Sun, Tseng-Yao and Kyoung S. Park, "Mechanical/Electrical." *Architectural Graphic Standards.* New York: John Wiley & Sons, Inc., 1971. Chaps. 15-16.

MECHANICAL SYSTEMS—REFRIGERANT: HEAT PUMP SYSTEM

A self-contained heat pump is a combined cooling and heating unit powered by electricity. The cooling cycle is identical to the basic compressive refrigeration cycle shown on page 97. However, during the heating cycle the flow of refrigerant can be reversed by a transfer valve to switch the heat exchange function of the evaporator (to a condenser) and condenser (to an evaporator). The most favorable heat pump applications are where the cooling and heating loads are almost equal. Heat pumps are often supplemented with electric resistance heaters for effective all-winter operation (e.g., outside coils may freeze).

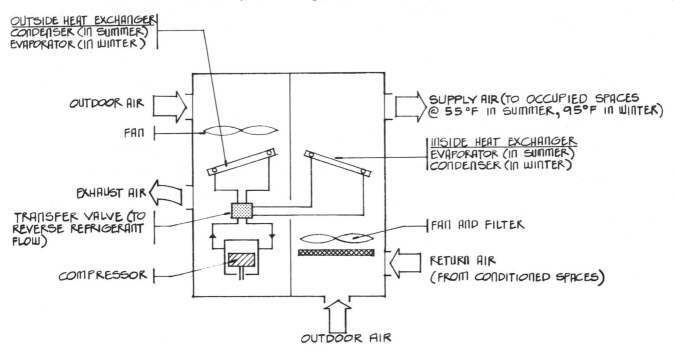

Note: When a balanced temperature demand condition exists, i.e., requiring all heating or all cooling, the heat pump principle is ideal. Unfortunately this condition rarely occurs in most buildings.

COMPARISON OF BASIC AIR-CONDITIONING SYSTEMS

Cost and comfort characteristics for the basic air-conditioning systems are shown below. It should also be noted that the specific application (e.g., apartment, motel, school, high-rise office) will often influence the system selection as availability and cost of energy, rental space requirements, extent of glazing, etc. can be overriding considerations.

System	Cost and Comfort Characteristics		
	Installation Cost	*Operating Cost*	*Comfort Performance**
Single duct	High	Medium	Medium
Double duct	High	High	High
Multi-zone	Medium	Medium	High
4- pipe	High	Medium	Medium
3- pipe	Medium	High	Low
2- pipe (Fan-coil)	Low	Medium	Low
2- pipe (Induction)	Medium	Medium	Medium
Packaged terminal air conditioner	Low	Medium	Medium

*Ranked on typical system performance in providing thermal comfort (i.e., desired air temperature, air motion, humidity) and occupant control on a year-round basis. For example, hot water (or air) and chilled water (or air) that is circulated through the same pipes (or ducts) gives occupants two choices: heating or nothing, and cooling or nothing. Therefore, during the changeable weather of spring and fall, 2-pipe systems can not provide full-time comfort. For a comprehensive discussion of rating systems for specific applications of air-conditioning systems, see: J. B. Olivieri, "Proposed Air-conditioning Rating System," ASHRAE Journal, January 1971.

ZONING TECHNIQUES

Zoning is the separation of the mechanical distribution system into a number of specific demand areas. Each zone is designed to maintain comfort conditions by a single set of controls as dictated by the outdoor climate conditions and the building activities. In addition, to conserve energy, zoning controls should be flexible so that only those spaces actually in use will be conditioned. Apartments, conference rooms, and auditoriums are examples of spaces that may be unoccupied for long periods of time. Note that some loads such as the heat gain from electrical-mechanical equipment usually will be constant throughout the building.

Low, Small Buildings

Larger heating units can be selected for the zone on the north exposure and larger cooling units for the zone on the south. However, where occupant activity is generally uniform throughout small buildings (e.g., see example duct sizing problem on page 128), the all-air, single duct system with thermostatically-controlled terminal reheat devices in the branch ducts can be used. These heating devices will allow the single duct system to respond to a moderate range of temperature demand within a given zone. In winter months, supplementary heat also can be furnished by fin-coil or electric resistance baseboard units located under large areas of exterior glass.

Tall Buildings

Zoning for tall buildings can be complex because solar radiation may vary significantly with elevation at various building exposures due to the shade provided by the surrounding structures. For example, on a cold, sunny day the side exposed without shade to the sun may require cooling whereas the shaded exposure requires heating. To respond to these simultaneous loads, tall buildings are often separated into south and east, north and west, and into vertical zones. Effective vertical zoning in tall buildings can also help neutralize the "stack effect" (i.e., the tendency of warm air to rise causing variations in building air static pressure). The stack effect can induce undesirable infiltration at low building levels and exfiltration at high levels. The all-air, double duct system, for example, can respond to a wide range of temperature demands by providing a supply of hot and cold air to each zone where it can be mixed according to specific demand requirements. For interior zones, the all-air, single duct variable volume system can conserve energy by varying the amount of air supplied to rooms based on specific cooling demand requirements.

MECHANICAL SYSTEMS—ZONING EXAMPLES

Horizontal air duct distribution networks can be located above suspended ceilings, under floors, and sometimes integrated with structural elements. Shown below are examples of a single duct system serving two zones per level and a multi-zone system serving five zones.

Single Duct System (Two Zones)

In the low-rise building shown below, conditioned air is distributed to separate north-west (number 1) and south-east (number 2) zones. In winter, supplementary under-glass heat can be provided by fin-coil or electric resistance devices.

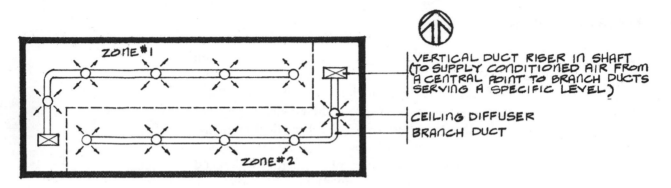

VERTICAL DUCT RISER IN SHAFT (TO SUPPLY CONDITIONED AIR FROM A CENTRAL POINT TO BRANCH DUCTS SERVING A SPECIFIC LEVEL)

CEILING DIFFUSER
BRANCH DUCT

Multi-Zone System (Five Zones Shown)

The all-air multi-zone system supplies mixed hot and cold air at low velocities to individual zones from a central equipment space. Room thermostats control a pair of dampers at the supply unit to proportion the mixture of conditioned air delivered to each zone. Horizontal duct runs are typically 100 to 200 ft in length.

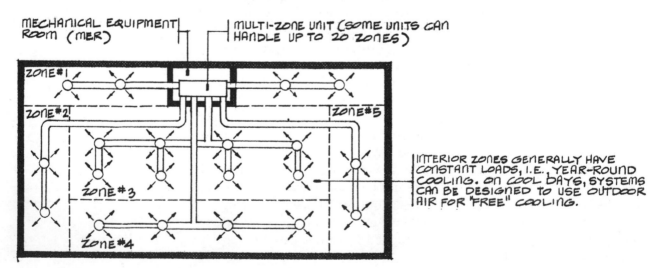

MECHANICAL EQUIPMENT ROOM (MER)

MULTI-ZONE UNIT (SOME UNITS CAN HANDLE UP TO 20 ZONES)

INTERIOR ZONES GENERALLY HAVE CONSTANT LOADS, I.E., YEAR-ROUND COOLING. ON COOL DAYS, SYSTEMS CAN BE DESIGNED TO USE OUTDOOR AIR FOR "FREE" COOLING.

Note: The mechanical space for multi-zone systems should be in a central location to minimize the length of duct runs. Systems having one central fan can *not* operate when the fan unit requires maintenance or repair service.

MECHANICAL SYSTEMS—AIR DUCT SYSTEMS FOR CONCRETE SLABS

Typical single zone duct system layouts for warm-air furnace units, used in residences, schools, small retail shops and stores, etc., are shown below. The basic types are radial, perimeter loop, and lateral systems. They are generally installed in the floor slab. Consequently, care must be exercised when pouring the concrete to avoid damaging or dislodging ductwork.

Radial

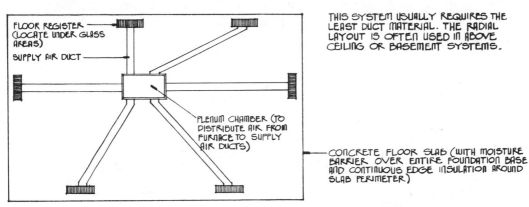

THIS SYSTEM USUALLY REQUIRES THE LEAST DUCT MATERIAL. THE RADIAL LAYOUT IS OFTEN USED IN ABOVE CEILING OR BASEMENT SYSTEMS.

FLOOR REGISTER (LOCATE UNDER GLASS AREAS)

SUPPLY AIR DUCT

PLENUM CHAMBER (TO DISTRIBUTE AIR FROM FURNACE TO SUPPLY AIR DUCTS)

CONCRETE FLOOR SLAB (WITH MOISTURE BARRIER OVER ENTIRE FOUNDATION BASE AND CONTINUOUS EDGE INSULATION AROUND SLAB PERIMETER)

Perimeter Loop

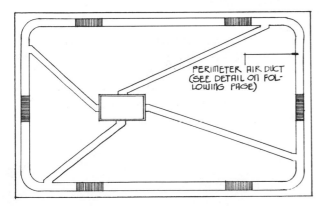

THE PERIMETER LOOP DUCT REDUCES SLAB EDGE HEAT LOSSES CONSIDERABLY. THE PLENUM CHAMBER IS OFTEN CENTRALLY LOCATED IN SLAB SYSTEMS BUT MAY ALSO BE PLACED NEAR THE SLAB EDGE OVER PERIMETER LOOP DUCTS IN THE LOOP AND LATERAL SYSTEMS.

PERIMETER AIR DUCT (SEE DETAIL ON FOLLOWING PAGE)

Lateral

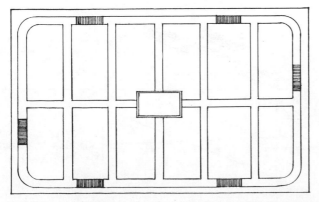

THE LATERAL SYSTEM USES THE GREATEST AMOUNT OF DUCT MATERIAL BUT ACTS AS A RADIANT PANEL PROVIDING MORE UNIFORM SLAB TEMPERATURES.

Note: Duct systems also may be installed below ground level spaces in a basement or crawl space. Be sure ductwork is insulated and airtight to reduce heat loss.

MECHANICAL SYSTEMS—STRUCTURAL-MECHANICAL DUCT-WORK

Air ducts in all-air and air-water air-conditioning systems take up considerable space. Consequently, integration of the air paths within (or between) structural elements can conserve important useable area and volume.

Tee Beams

The space between the stems of adjacent tee beams can be used for an air path as shown below. Openings for vertical duct connections, floor registers, etc. can be made during fabrication of the beams or drilled in the field.

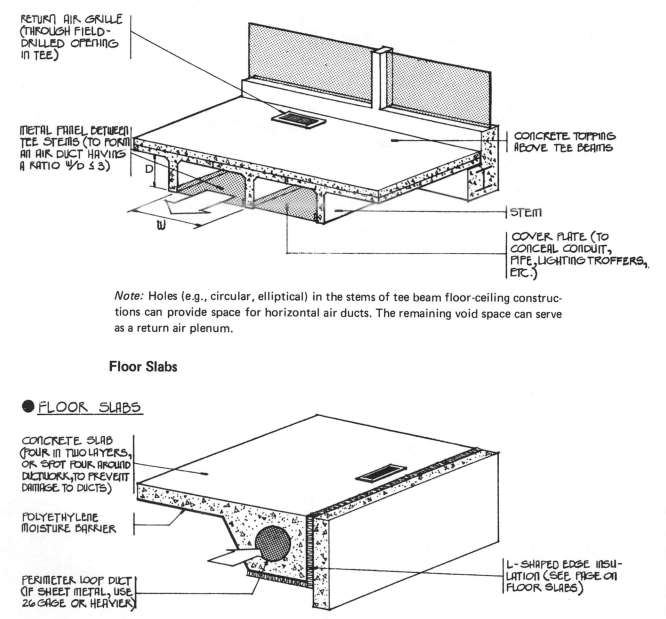

Note: Holes (e.g., circular, elliptical) in the stems of tee beam floor-ceiling constructions can provide space for horizontal air ducts. The remaining void space can serve as a return air plenum.

Floor Slabs

EQUIPMENT PLANT LOCATION ALTERNATIVES

The best location in a tall building for a central mechanical equipment space depends on many factors. Some of the major considerations for evaluating roof and basement plant locations are given below. Fan rooms, requiring about 2 to 4% of total building floor area, generally are located to serve specific zones or levels. One fan room can serve about eight to twenty floors. However, the fewer floors served by a fan room, the smaller the vertical duct shaft and fan room height requirements. A typical distribution layout for fan rooms in tall commercial buildings is to downfeed air for the top third of the building from a roof location and to upfeed the middle third and downfeed the lower third from a lower third point location. Fan room location also should be convenient to the outdoors for air intake and exhaust purposes. Note that vertical air duct shaft space is usually about 4 sq ft for every 1,000 sq ft of floor area served. Vertical duct shafts can be located in a central core area or at the ends or corners of a tall building. Typically, the maximum area served by a shaft can be estimated by a 65 ft radius from the shaft.

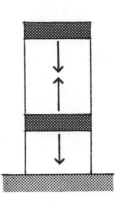

Roof Location

The major advantages of a roof (or "penthouse") location are: shorter condenser and chilled water lines as cooling tower and most other equipment will be together, elimination of boiler flue shaft space requirements, and no loss of rental space or below-grade parking. Disadvantages are: structural loads on the roof are increased, noise and vibration control design is more complicated, maintenance is difficult, and replacement of equipment requires hoisting operations. Note that cooling towers usually require about 1 sq ft of roof area for every 400 sq ft of floor area served.

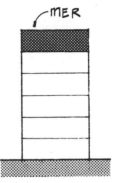

Basement Location

The major advantages of locating the mechanical plant below-grade are: reduced structural loads, relative ease of equipment maintenance and replacement, and potential for early occupancy of lower level spaces if fans are used at each floor level. Disadvantages are: requirements for providing plant ventilation below-grade, lost basement rental income, and lost space and potential noise and vibration problems from pipe risers (e.g., water circuit between condenser in basement and cooling tower located on roof). Note that air intake grille (or louver) openings to replenish air consumed by burners and to ventilate the mechanical space can be sized at 1 sq in. per 1,000 Btuh for heating equipment (e.g., boilers, water heaters, incinerators), at 1 sq in. per 2,000 Btuh if forced draft boilers are used. For noise control, carefully locate all openings to the outside and, if necessary, use sound attenuating air-vent devices.

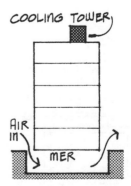

SPACE REQUIREMENTS FOR VARIOUS SYSTEMS

Realistic estimates of mechanical space requirements should be made at the initial design stages of a project so only minor changes will be dictated by the final mechanical system design. The mechanical equipment room (usually abbreviated MER) generally should be in a central location and have direct access to the outdoors. Inefficient mechanical room locations can cause considerable loss of building space. Any criss-crossing of air ducts, for example, will result in waste of useable area and volume. Central locations, that consolidate major equipment in unoccupied areas, can shorten air duct lengths and prevent excessive duct heat loss or gain where they pass through unconditioned spaces. In addition, central plants generally will have larger, more efficient equipment. Some guidelines to help determine mechanical equipment room space requirements, clearance and floor area, for basic air-conditioning systems are given below.

Room Clearances

For preliminary layouts, provide 12 ft minimum clear height under the ceiling structure for commercial buildings, 14 ft for industrial buildings. The final clear height will be set by the type of heating and refrigeration plants used and by the actual equipment placement (e.g., major units on roof, in central room, or separate room on each floor). Be sure to allow clearances for tube and coil removal (e.g., boilers, chillers, fans, etc.) and to provide adequate access for maintenance of filters, traps, and similar components.

Floor Areas

The percentages of gross floor area required for the mechanical space, indicated in the table, are average values based on central heating and refrigeration plants. If the building has an external central plant, for instance, the percentages should be reduced. The sketches at the left show the vertical shaft space required for equal cooling capacity from an example all-air and air-water system.

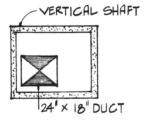

VERTICAL SHAFT

24" x 18" DUCT

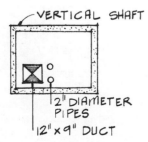

VERTICAL SHAFT

2" DIAMETER PIPES

12" x 9" DUCT

System	Example	Floor Area Required For Mechanical Space (%)*
All-Air**	Single duct (terminal reheat)	4 to 8
	Single duct (variable volume)	3 to 6
	Double duct	4 to 8
All-Water	4- pipe	1.5 to 5
	3- pipe	1.5 to 5
	2- pipe	1 to 3
Air-Water**	Induction (2- pipe)	3 to 6
Refrigerant	PTAC	1 to 3
	Roof-top	0 to 1

*Use upper end of percentage range for smaller buildings — less than about 10,000 sq ft — and for buildings having heavy conditioning loads (e.g., laboratories, hospitals, auditoriums).

**Separate rooms for air-handling equipment will require 2 to 4% of total occupied area.

EQUIPMENT WEIGHTS FOR STRUCTURAL LOADING

The equipment weight factors given below are based on averages of data from various manufacturers. They may be used to estimate the installed equipment structural loadings as the data includes allowance for the weight of accessory equipment such as controls, piping, and supports. Try to locate heavy equipment close to structural columns and load bearing walls. Do *not* place roof-top equipment near mid-span on light-weight frame structures.

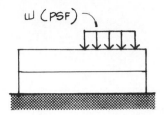

Heating	*Lbs (per 1,000 Btuh)*
Boiler	7 to 2
Warm-air furnace	5 to 2
Cooling	*Lbs (per ton of refrigeration)*
Chiller, absorption	120 to 60
Chiller, centrifugal	120 to 40
Compressor	100 to 40
Condenser	175 to 50
Cooling tower	100 to 50
Air-handling	*Lbs (per 1,000 cfm)*
Fan	200 to 30
Fan-coil	200 to 40
Miscellaneous	*Lbs (per ton of refrigeration)*
Heat pump	160 to 120
PTAC	200 to 140
Pump	10 to 5

The operating weight of the equipment specified for a project should be furnished by the manufacturer. For preliminary estimates, however, use the data above or a live load in the range of 125 to 225 pounds per square foot (psf) for the mechanical space.

DUCT SIZING PROCEDURE

In the velocity method for sizing ductwork, an air velocity is assigned to each section of the duct system subject to certain preferred ranges based on noise control considerations. The velocity should be highest at the fan outlet and reduced as various branches are taken off the main duct. Since the airflow volume in cubic feet per minute (cfm) for each section is also known, assuming the air velocity in feet per minute (fpm) allows direct calculation of the required duct sizes. The step-by-step supply duct sizing procedure for small, low velocity systems (e.g., low-rise offices, residences, retail shops and stores) is outlined below. These procedures, however, are valid in principle for other basic air-conditioning systems such as double duct, induction, etc.

Layout

Locate the mechanical equipment room (MER) in a central area adjacent to an outside wall for ease in equipment servicing as well as for access to outdoor air. Prepare an initial duct system layout connecting the fan in the MER to the supply air outlets in each room as dictated by the building configuration (i.e., obstructions, clearances, architectural considerations). Avoid complex duct layouts and any crisscrossing of ducts. If duct crossovers are required, however, try to utilize the space between beams. Be careful to locate the supply air outlets observing the basic principles of room air motion. Generally, ceiling diffusers are preferred where room airflow volumes exceed 2 cfm/sq ft of floor area. Label each section of duct in some convenient fashion starting at the MER.

Airflow Volume

Determine the required airflow volume (Q) in cfm for each room based on the largest value of cfm demand from:

1. Heat loss analysis, or

2. Heat gain analysis, or

3. Outdoor air requirement

If more than one register is used in a room, divide the total cfm among the registers and note on the initial duct system layout. Write the air quantity on each section of duct starting at the register located farthest from the MER, working backward to the MER. Avoid abrupt turns in duct layouts to achieve minimum pressure losses and to prevent noise generation.

Air Velocity

Assume an air velocity (v) for each duct section based on tables of recommended air velocities for duct systems (e.g., see page 120).

DUCT SIZING PROCEDURE (Continued)

Cross-Section Area

Calculate the area of each duct section using the formula:

$$A = 144 \frac{Q}{v}$$

where A = cross-section area in sq in.

         Q = airflow volume in cfm

         v = air velocity in fpm

Next, find the diameter (d) of a round duct having this area (A) as friction loss tables and graphs are usually based on circular sections.

$$d = \sqrt{1.3A}$$

where d = diameter of round duct in inches

         A = cross-section area in sq in.

Convert the diameter to an equivalent rectangular duct using the tables for equal friction and capacity. Since resistance to airflow by friction depends on surface area, *not* cross-section area, round ducts offer the least resistance. However, they take up more space than equivalent rectangular ducts. The ratio of the long side to the short side of a straight duct (often referred to as the "aspect ratio") should not exceed 5:1 although a 4 or 3:1 limit is preferred. For example, a 60 by 12 duct has a cross-section area equal to that of a 30 by 24 duct, but its perimeter is 3 ft greater. In addition, the 60 by 12 duct in this example requires more material, meaning increased ductwork cost.

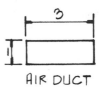

AIR DUCT

Ducts are insulated to control condensation and heat loss (or gain). Allow at least a 2 in. increase in overall duct dimensions of width and depth on the drawings if internal duct insulation is used for noise control purposes. Always specify duct dimensions in inches on mechanical system drawings. Be careful as duct sizes are often given as clear inside dimensions on drawings.

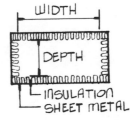

System Pressure Drop

Calculate the pressure drop in inches of water (abbreviated "w.g., i.e., inches water gauge) by summing the friction losses of the individual sections making up the longest total length (actual plus equivalent length). Note that a column of water 1 in. high exerts a pressure of 0.04 psi at its bottom.

    1.   *Elbow losses:* Use the graph on page 124 to determine equivalent length from elbows. Add this additional length figure in feet to the actual length of the duct section in which it occurs.

2. *Duct friction:* Use the friction chart for air. Friction loss in ducts increases with air velocity, interior surface roughness and duct size and length. The chart relates air quantity in cfm, velocity in fpm, and duct diameter in inches to a corresponding friction loss in inches of water *per 100 ft* of duct length. Find the loss for each section in longest run.

3. *Branch take-off losses:* Use the graph on page 125 to find pressure losses for 90° and angular take-offs.

4. *Dynamic losses:* Whenever the velocity of airflow in a duct is changed, the pressure loss will be greater than if there had been uninterrupted flow. This additional loss is usually termed "dynamic loss." The dynamic losses at changes in cross-section due to duct transitions can be neglected in low velocity systems if the transition is smooth, i.e., 45° or less. To simplify duct installation, duct transitions often can be achieved by a change in only one dimension—preferably the width.

5. *Register losses:* Consult manufacturer's catalogs for pressure loss of supply air outlet unit selected. In low velocity systems, for example, diffuser pressure losses can vary from about 0.01 to 0.10 "w.g.

Add the pressure losses from steps 1 through 5 above. The pressure capabilities of air-handling units should be obtained from the manufacturer as total system pressure drop must not exceed these figures. Dampers can be used to "balance" pressure drops within duct networks by creating equal pressure drops in the other duct sections.

Register Noise

Air velocities at room supply outlets should not exceed the values presented in the table on page 144. If specific acoustical data for manufacturer's equipment is available, check this against the desired acoustical background levels.

Note: Two additional methods ("equal friction" and "static regain") for designing ductwork are also available. For detailed information on these methods, see: Trane Air Conditioning Manual (1969) or Chapter 25, ASHRAE Guide and Data Book: Fundamentals (1972).

OUTDOOR AIR REQUIREMENTS

Type of Space*	Outdoor Air (cfm/occupant)
Auditoriums, theaters, retail shops and stores, beauty parlors, etc.	5 to 15
Restaurants, banks, barber shops, etc.	10 to 15
Offices, reception areas, hospitals**, apartments, laboratories, etc.	10 to 30
Hotels, motels, conference rooms, etc.	25 to 50

*For other general applications, where room occupants can be expected to smoke cigarettes and cigars, use a range of 25 to 40 cfm per occupant. Where it is absolutely certain that smoking will *not* be permitted, use at least 5 to 10 cfm per occupant.

**Clean, odor-free outdoor air is essential to overcome the explosion hazard of anesthetics in operating rooms. Minimum requirements for safe practice are given in the National Fire Protection Association Pamphlet No. 56A. (Available from the National Fire Protection Association (NFPA), 470 Atlantic Avenue, Boston, Massachusetts 02210.)

Source: ASHRAE Handbook of Fundamentals (1972), p. 421.

RECOMMENDED AIR VELOCITIES FOR DUCT SYSTEMS

Location	Preferred Range of Air Velocities (fpm)
Outdoor air intake louvers*	250 to 500
Filters*	250 to 350
Heating and cooling coils*	450 to 600
Fan outlets	1,000 to 2,400
Main ducts**	700 to 1,200
Branch ducts**	450 to 1,000
Supply registers and return grilles	(See page 144.)

*Air velocities are given for the face area, whereas, other velocities in table apply to net open (or "free") area.

**Air velocities are specified for low velocity duct systems only. It is good practice to use values from the low end of the preferred velocity ranges where quiet acoustical background levels are required, and where airflow volumes are low (less than 1,000 cfm). Conversely, where higher backgrounds are acceptable, values from the upper end may be used.

CIRCULAR EQUIVALENTS FOR RECTANGULAR DUCTS OF EQUAL FRICTION AND CAPACITY

The table lists circular ducts (diameter in inches) which, at the same air-flow volume, will have friction losses identical to the given rectangular ducts. The sides of the rectangular ducts are listed in inches in the column at the left and across the top row. For example, a 16 in. diameter round duct (located in body of table) is equivalent to an 18 by 12 rectangular duct. When a duct's diameter falls between values in the columns, use the higher value to define the side dimensions.

Side	4	6	8	10	12	18	24	30	36	42	48	60
3	3.8	4.6	5.2									
4	4.4	5.3	6.1	6.8	7.3							
5	4.9	6.0	6.9	7.6	8.3							
6	5.3	6.6	7.5	8.4	9.1	11.0						
7	5.7	7.1	8.2	9.1	9.9	11.9						
8	6.1	7.5	8.8	9.8	10.8	12.9	14.6					
9	6.4	8.0	9.3	10.4	11.3	13.7	15.6					
10	6.8	8.4	9.8	10.9	11.9	14.5	16.6					
12	7.3	9.1	10.8	11.9	13.1	16.0	18.3					
18	--	11.0	12.9	14.5	16.0	19.7	22.6					
24	--	--	14.6	16.6	18.3	22.6	26.2					
30	--	--	--	18.3	20.2	25.2	29.3	32.8				
36	--	--	--	--	21.9	27.4	32.0	35.8	39.4			
42	--	--	--	--	--	29.4	34.4	38.6	42.4	45.9		
48	--	--	--	--	--	31.2	36.6	41.2	45.2	48.9	52.6	
60	--	--	--	--	--	--	40.4	45.8	50.4	54.6	58.5	65.7
72	--	--	--	--	--	--	43.8	49.7	54.9	59.6	63.9	71.7
84	--	--	--	--	--	--	--	53.2	58.9	64.1	68.8	77.2

Note: To find duct sizes not given in the above shortened table, see ASHRAE Guide and Data Book: Fundamentals (1972) or Trane Air Conditioning Manual (1969).

MECHANICAL SYSTEMS—USE OF AIR FRICTION CHART

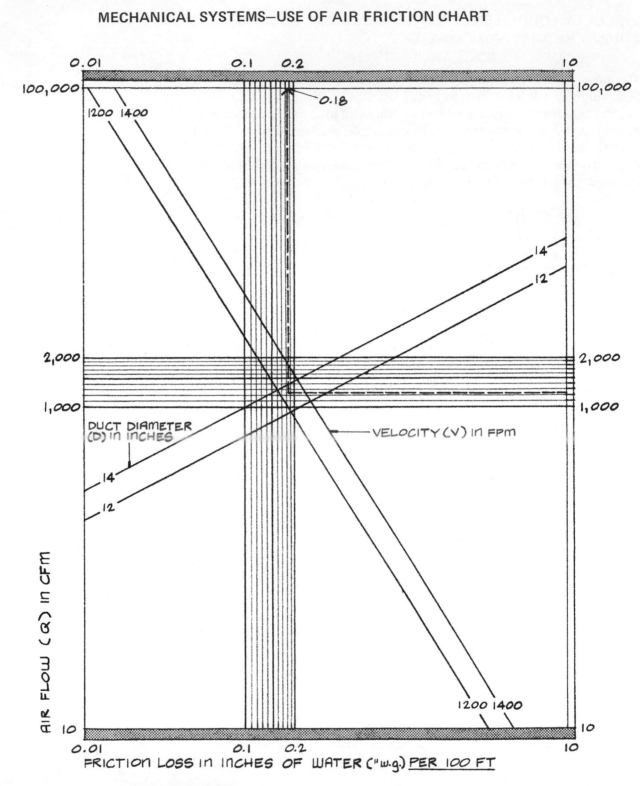

Example—Use of Chart

Given: Duct velocity (v) is 1250 fpm and airflow (Q) is 1262 cfm.

Procedure to find friction loss: Enter friction chart at Q = 1262 cfm and read opposite v = 1250 fpm curve above (or below) to a friction loss of 0.18 ''w.g. per 100 ft (note also that the chart can be used to find approximate duct diameter, e.g., about 14 in. for this example).

MECHANICAL SYSTEMS—AIR FRICTION CHART

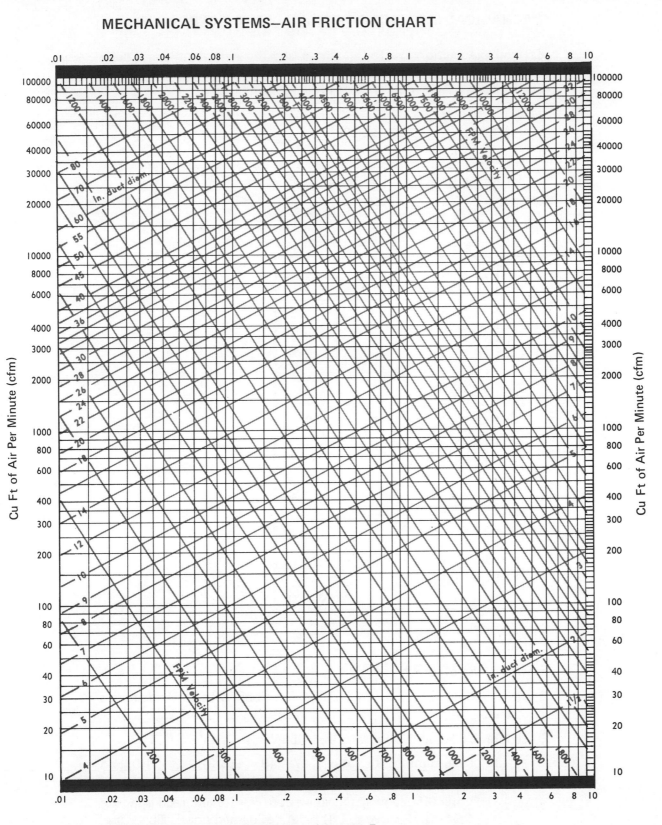

Friction Loss in Inches of Water ("w.g.) Per 100 Ft

Based on Standard Air of 0.075 lb per cu ft density flowing thru average, clean, round, galvanized metal ducts having approximately 40 joints per 100 ft. For losses in flexible duct and factory-made spiral-wound metal duct, see data published by manufacturer.   COURTESY OF ASHRAE

MECHANICAL SYSTEMS—FRICTION LOSS FOR ELBOWS (90°)

To find the friction loss for an elbow, add the equivalent length determined by the graph below to the actual length of duct section where it occurs. Use the air friction chart to find loss in "w.g. per 100 ft.

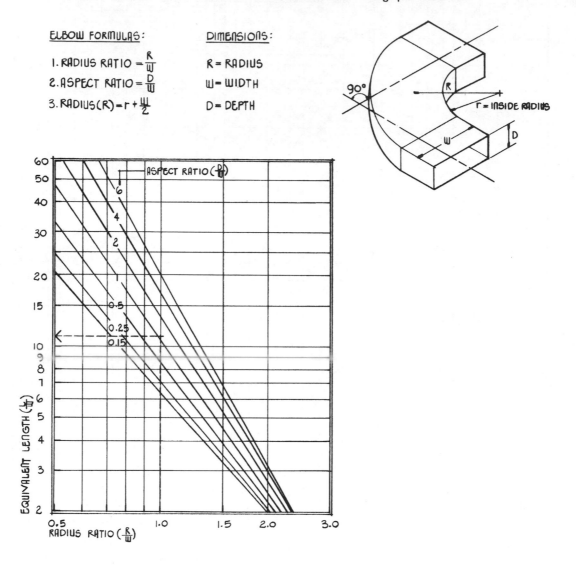

ELBOW FORMULAS:

1. RADIUS RATIO = $\frac{R}{W}$
2. ASPECT RATIO = $\frac{D}{W}$
3. RADIUS (R) = $r + \frac{W}{2}$

DIMENSIONS:

R = RADIUS
W = WIDTH
D = DEPTH

r = INSIDE RADIUS

Example—Use of Graph

Given: 18 in. wide (W) by 24 in. deep (D) elbow with an inside radius (r) of 9 in.

Procedure to find equivalent length of elbow:

1. Determine radius ratio, R/W = [9 + (18/2)] /18 = 1.0; and aspect ratio, D/W = 24/18 = 1.33.
2. Enter 90° elbow graph above at R/W = 1.0 and read opposite D/W = 1.33 curve to L/W ≃ 12 (See dashed lines). Therefore, elbow equivalent length is L = 12 W = 12 x 18 = 216 in. or 216/12 = 18 ft.

Note: This graph is also valid for elbows less than 90°. For example, 45° elbows have about $\frac{1}{2}$ the friction loss of 90° elbows, 30° elbows about $\frac{1}{3}$, etc.

BRANCH RATIO FORMULA:

$$R = Q_2 / Q_1$$

WHERE;

Q_2 = BRANCH AIR FLOW IN CFM

Q_1 = MAIN DUCT AIR FLOW UPSTREAM FROM BRANCH IN CFM

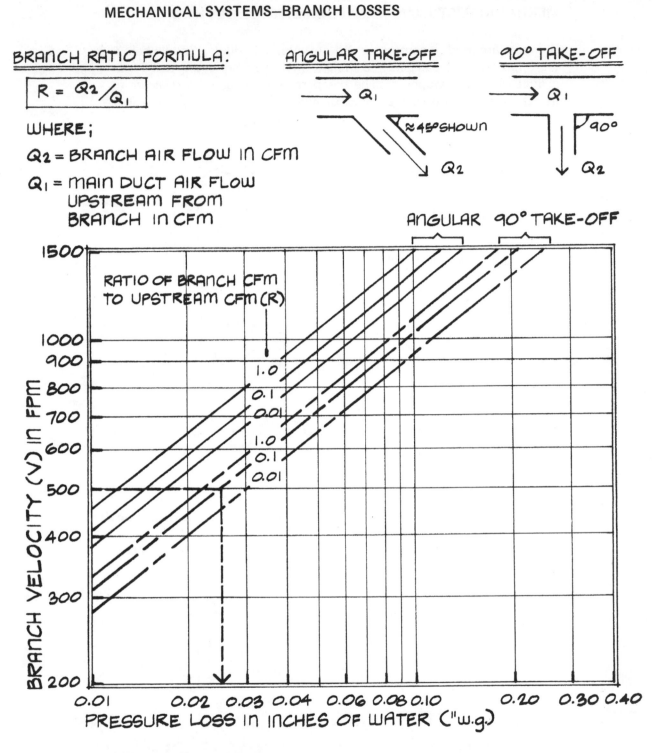

Example—Use of Graph

Given: Main duct airflow (Q_1) upstream from 90° take-off is 8000 cfm and branch airflow (Q_2) is 800 cfm. Branch velocity is 500 fpm.

Procedure to find branch pressure loss:

1. Calculate $R = Q_2/Q_1 = 800/8000 = 0.1$.
2. Enter graph at v = 500 fpm and read opposite R = 0.1 curve to pressure loss of 0.025 "w.g. (See dashed lines on graph.)

MECHANICAL SYSTEMS—AIRFLOW CONTROL DEVICES

Dampers are used to control the airflow volume in ducts. Shown below are opposed-blade dampers in a straight duct and a deflecting damper at a duct division. Opposed-blade dampers are also used in registers to throttle airflow into rooms. Most dampers are operated by removable keys or levers.

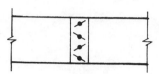

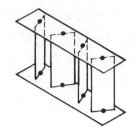

OPPOSED-BLADE DAMPERS

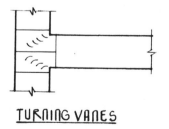

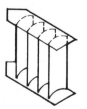

DEFLECTING DAMPER

Note: Fire dampers that are used where ducts penetrate fire walls, at air outlet devices, etc. will be specified by local building codes. They usually operate by a fusible-link set to close at a specific temperature.

Turning vanes and splitters help reduce air turbulence at elbows (and tees) by distributing airflow evenly across the duct cross-section. Generally, elbows having a radius ratio (*R/W*) less than about 1.0 always require turning vanes or splitters. Note that a small radius ratio means sharp, abrupt turns.

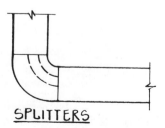

TURNING VANES

SPLITTERS

Note: Perforated, double-radius turning vanes are commercially available that enclose a sound-absorbing material to reduce airflow noise at elbows and tees.

MECHANICAL SYSTEMS—DUCTWORK SYMBOLS

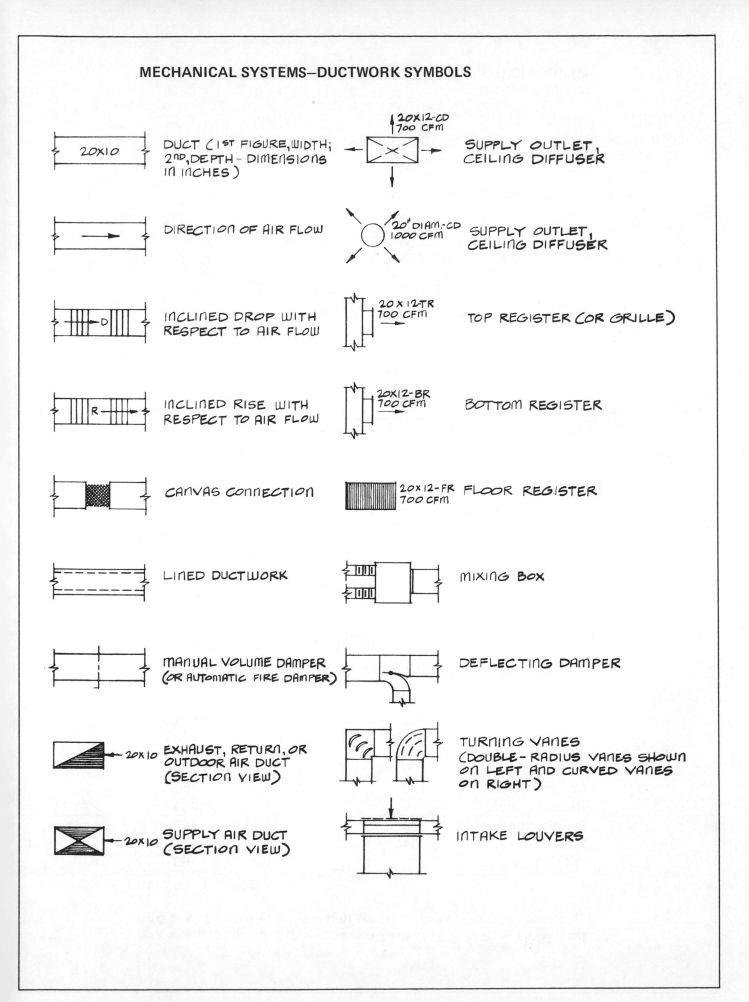

20X10	DUCT (1ST FIGURE, WIDTH; 2ND, DEPTH – DIMENSIONS IN INCHES)
→	DIRECTION OF AIR FLOW
D	INCLINED DROP WITH RESPECT TO AIR FLOW
R	INCLINED RISE WITH RESPECT TO AIR FLOW
	CANVAS CONNECTION
	LINED DUCTWORK
	MANUAL VOLUME DAMPER (OR AUTOMATIC FIRE DAMPER)
20X10	EXHAUST, RETURN, OR OUTDOOR AIR DUCT (SECTION VIEW)
20X10	SUPPLY AIR DUCT (SECTION VIEW)

20X12-CD 700 CFM	SUPPLY OUTLET, CEILING DIFFUSER
20" DIAM.-CD 1000 CFM	SUPPLY OUTLET, CEILING DIFFUSER
20 X 12-TR 700 CFM	TOP REGISTER (OR GRILLE)
20X12-BR 700 CFM	BOTTOM REGISTER
20X12-FR 700 CFM	FLOOR REGISTER
	MIXING BOX
	DEFLECTING DAMPER
	TURNING VANES (DOUBLE-RADIUS VANES SHOWN ON LEFT AND CURVED VANES ON RIGHT)
	INTAKE LOUVERS

MECHANICAL SYSTEMS—EXAMPLE PROBLEM: DUCT SYSTEM SIZING

GIVEN:

ARCHITECT/ENGINEER'S OFFICE, LOCATED IN NEW ORLEANS, LA, HAS THE FOLLOWING DESIGN CONDITIONS:

SUMMER — 95°F OUTDOOR DRY BULB TEMPERATURE (DB)

      53% RELATIVE HUMIDITY (RH)

      80°F OUTDOOR WET BULB TEMPERATURE (WB)
      (USE PSYCHROMETRIC CHART AS WB
      TEMPERATURE IS NEEDED TO SIZE
      COOLING AND OTHER EQUIPMENT.)

WINTER — 25°F OUTDOOR DRY BULB TEMPERATURE

INDOOR — 75°F DB AND 50% RH (SUMMER)

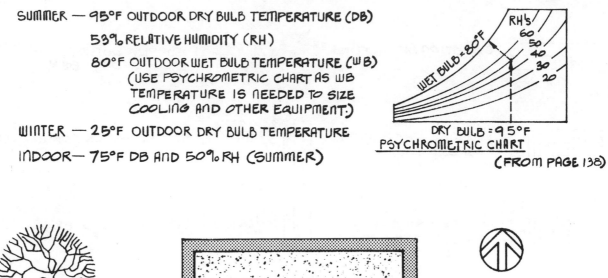

PSYCHROMETRIC CHART

(FROM PAGE 138)

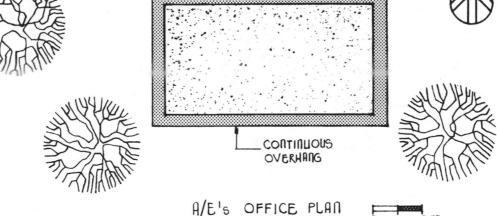

CONTINUOUS OVERHANG

A/E's OFFICE PLAN

0 5 10 FT

The space requirements and design Q's in *cfm* are given in the table below:

Space	Floor Area (ft²)	Occupants	Supply Q (cfm)
Private office	120	2	180
Conference room	120	4	180
Reception area	216	3	300
Drafting room	288	3	440
MER	64*	—	—
W.C.	40	—	60
			Total: 1,160 *cfm*

*Mechanical equipment room (MER) has 10 ft clear height and with 7% of gross floor area figure for all-air systems, will require 0.07 × 24 × 40 ≃ 67 sq ft floor area.

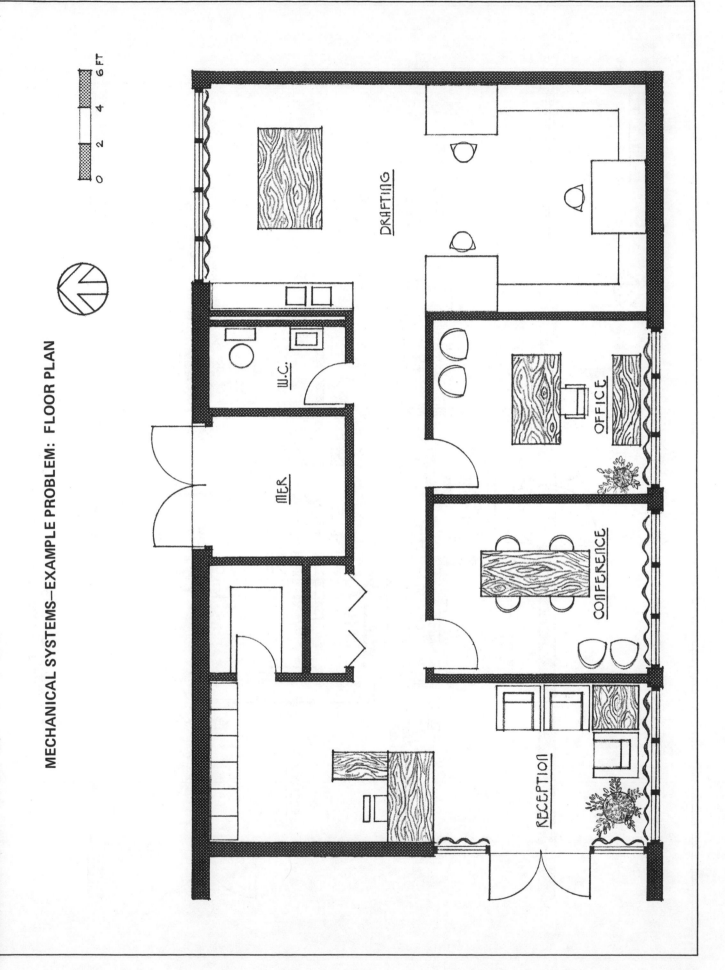

MECHANICAL SYSTEMS—EXAMPLE PROBLEM: FLOOR PLAN

DRAFTING

W.C.

MER

OFFICE

CONFERENCE

RECEPTION

0 2 4 6 FT

MECHANICAL SYSTEMS—EXAMPLE PROBLEM: DUCT SYSTEM LAYOUT

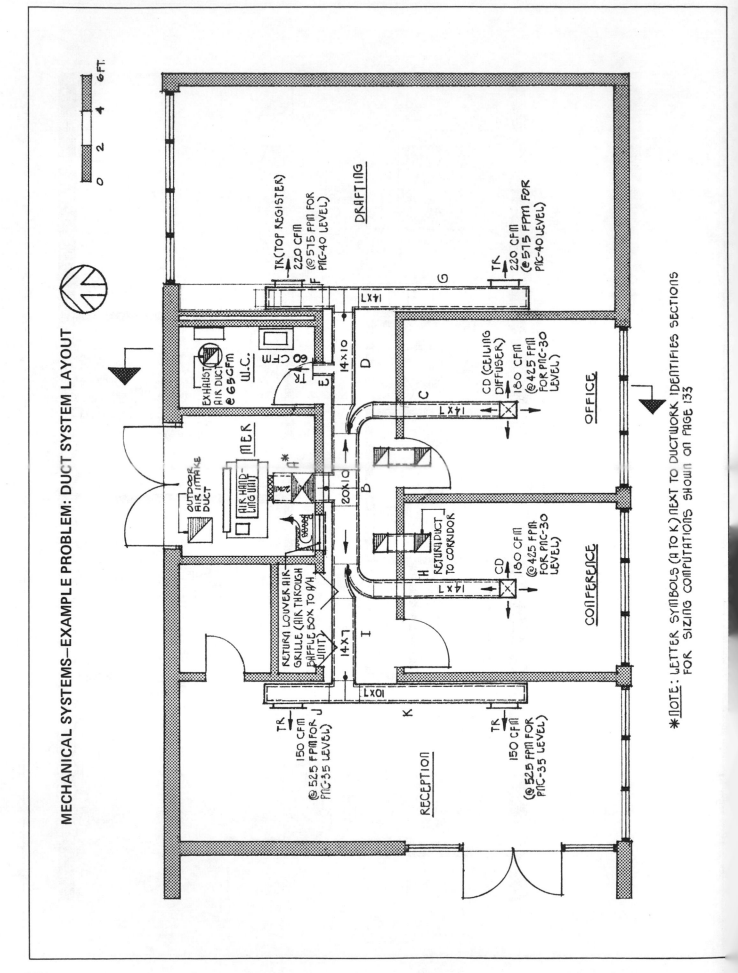

MECHANICAL SYSTEMS—EXAMPLE PROBLEM: DUCT SYSTEM DETAILS

Part of the example problem duct system is shown in sectional perspective below. See plan view on preceding page for location of the section.

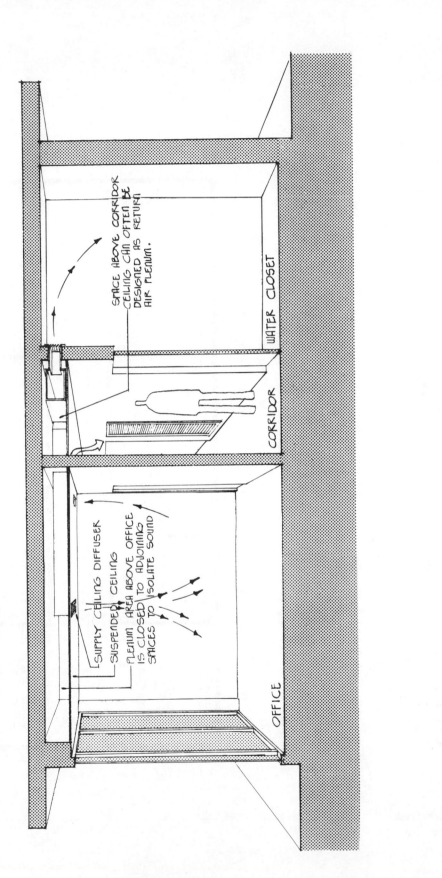

SPACE ABOVE CORRIDOR CEILING CAN OFTEN BE DESIGNED AS RETURN AIR PLENUM.

WATER CLOSET

CORRIDOR

SUPPLY CEILING DIFFUSER
SUSPENDED CEILING
PLENUM AREA ABOVE OFFICE IS CLOSED TO ADJOINING SPACES TO ISOLATE SOUND

OFFICE

MECHANICAL SYSTEMS—EXAMPLE PROBLEM: INITIAL SUPPLY SYSTEM DUCT LAYOUT

Procedure to layout duct network:

1. Indicate air outlet position (symbol ⊨) and cfm on plan drawing. Required total cfm for room will be its maximum value of: heat loss/gain analyses or outdoor air requirements.

2. Connect fan to air outlets. Branch ducts extend from large main duct in same manner branches extend from a tree trunk.

3. Determine cfm for each section of duct network by summing total cfm of outlets served by that section. (See cfm values adjacent to letter section identifications.)

4. Air velocities (shown in parentheses) are selected from table on page 120.

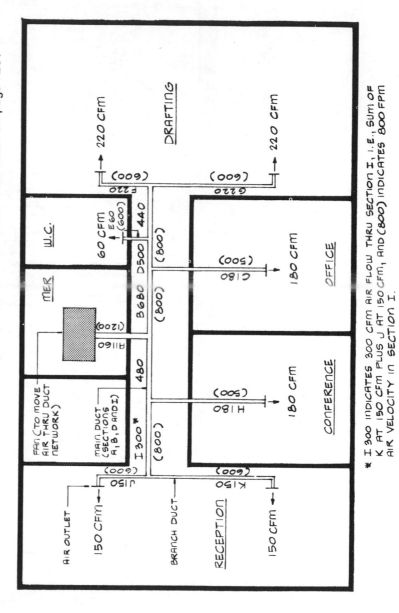

* I 300 indicates 300 cfm air flow thru section I, i.e., sum of K at 150 cfm plus J at 150 cfm, and (800) indicates 800 fpm air velocity in section I.

Note: Required duct sizes are shown in table on following page.

DUCT SIZING TABLE FOR THE EXAMPLE PROBLEM

Section	v (fpm)	Q (cfm)	Area (in.^{2}) = 144 Q/v	Diameter (in.) = $\sqrt{1.3A}$	W X D*
A	1,200	1,160	140	13.5	18 X 9
B	800	680	122	12.6	18 X 8
C	500	180	52	8.2	12 X 5
D	800	500	90	10.8	12 X 8
E	600	60	14	4.3	4 X 4
F	600	220	53	8.3	12 X 5
G	600	220	53	8.3	12 X 5
H	500	180	52	8.2	12 X 5
I	800	300	54	8.3	12 X 5
J	600	150	36	6.9	8 X 5
K	600	150	36	6.9	8 X 5

*Duct dimensions of width (W) and depth (D) are given by circular equivalent table on page 121. Increase duct dimensions by 2 in. for inside 1 in. thick glass fiberboard lining to control condensation and duct-borne noise. Section B, for example, will become 20 by 10 as shown on the duct system layout drawing. Where ducts are internally lined, increase friction loss from chart on page 123 by 30% for air velocities of 450 to 850 fpm, 35% for 850 to 1,200 fpm.

Note: For rectangular sheet metal ducts having a long dimension of up to 12 in., use 26 gage steel; for 13 to 30 in.—24 gage; and for 31 to 54 in.—22 gage.

PRESSURE DROP TABLE FOR THE EXAMPLE PROBLEM

Item	Actual or Equivalent Length (ft)	Friction Loss ("w.g./100 ft)	Chart Ratios and Friction Computations	Pressure Drop ("w.g.)*
A's Elbows	26	0.17	$\dfrac{D}{W} = \dfrac{9}{18} = 0.50$ and $\dfrac{R}{W} = 1.0$; $L = W(8.5) = \dfrac{18(8.5)}{12} = 13$ ft each (See page 124 for an elbow friction loss example solution.)	—
A's Tee**	13	0.17	ditto	—
A	6***	0.17	$\dfrac{0.17\,(26 + 13 + 6)}{100} \times 1.35$	0.103
B	4	0.08	$\dfrac{0.08\,(4)}{100} \times 1.30$	0.004
90° Take-off	—	—	$R = Q_2/Q_1 = \dfrac{180}{680} = 0.3$ and $V = 500$ fpm (See page 125 for an example problem showing branch losses at take-offs.)	0.031
D's Tee**	9	0.10	$\dfrac{D}{W} = \dfrac{8}{12} = 0.67$ and $\dfrac{R}{W} = 1.0$; $L = W(9) = \dfrac{12(9)}{12} = 9$ ft	—
D	7	0.10	$\dfrac{0.10\,(9 + 7)}{100} \times 1.30$	0.021
90° Take-off	—	—	$R = Q_2/Q_1 = \dfrac{60}{500} = 0.12$ and $v = 600$ fpm	0.046
G	9	0.08	$\dfrac{0.08\,(9)}{100} \times 1.30$	0.009
Register	—	—	(See manufacturer's data for loss at register's operating conditions.)	0.060
			Total Supply System Pressure Loss:	0.274"w.g.

*Pressure drop computed for individual supply system sections making up the longest total length (actual plus equivalent length). Use air friction chart on page 123 to find loss in "w.g./100 ft.

**Consider pressure loss of tee equal to similar elbow based on the entering velocity.

***The straight length of duct between elbows (or tees) is measured to the intersection of the center lines of the straight duct sections.

NOTES ON THE EXAMPLE PROBLEM

Supply Air System

The supply air system duct layout is shown on page 130. Building air-flow volumes (Q) are determined by the largest demand from heat loss analysis, heat gain analysis, or outdoor air requirement on a room-by-room basis. Design Q's are shown at room air outlet devices on the layout drawing. Duct sizes and system pressure losses are summarized by the tables on pages 133 and 134. The air velocities at room registers (and diffusers) recommended for acceptable acoustical background noise levels are also indicated on the layout drawing. The noise levels in the private office and conference room should be about PNC-or NC-30* while the drafting room and reception area can be higher at PNC-40 and PNC-35 respectively as these spaces only require fair listening conditions. See page 144 for suggested air velocities at room registers for various spaces.

Outdoor Air for Odor Control

Outdoor air intake is based on a ventilation rate of about one building air change per hour (170 cfm) plus replacement of exhaust system air (65 cfm). Intake of 235 cfm will satisfy odor control requirements of 25 cfm per person for offices. Assuming a maximum occupancy of nine persons, the outdoor air intake should be at least 9 X 25 = 225 cfm. The quantity provided is $\frac{235}{1160} \simeq 20\%$ of building supply air system design Q. This percentage is generally referred to as replacement or "make-up" air.

Return Air System

Air from the private office and conference room is returned through separate internally lined transfer ducts in the plenum space above the suspended ceiling. Always avoid undercut door thresholds and louvered doors because they will provide poor acoustical privacy. Note that the air from the drafting room and reception area is returned to the air grille located in the corridor at the mechanical equipment room. This will be satisfactory because acoustical isolation between these two spaces is not critical. See page 147 for a discussion of return air systems.

Refrigeration Load

The psychrometric chart is used to size the cooling plant. Building sensible heat gain from solar radiation, lighting, etc. will be 25,000 Btuh. Heat from nine occupants, moderately active office workers (See page 81), can be estimated at 9 X 250 = 2,250 Btuh sensible heat gain and 9 X 200 = 1,800 Btuh latent heat gain (H_{lat}). Consequently, the ratio

---

*For a description of noise criteria (NC) levels appropriate for many activities, see: M. D. Egan, *Concepts in Architectural Acoustics.* New York: McGraw-Hill, Inc., 1972, p. 86.

NOTES ON THE EXAMPLE PROBLEM (Continued)

of sensible heat to total heat gain—called the sensible heat ratio (SHR)—will be

$$SHR = \frac{H_{sen}}{H_{sen} + H_{lat}} = \frac{27,250}{27,250 + 1,800} = 0.94$$

A step-by-step procedure to find the refrigeration load in tons is given on the following page. Note that enthalpy is the total heat, sensible and latent, of air and water vapor present in air in Btu/lb.

MECHANICAL SYSTEMS—REFRIGERATION LOAD ANALYSIS

Procedure for using psychrometric chart:

1. Locate point *A* (condition of room return air) at design indoor air temperature and *RH.*

2. Find point *B* (supply air condition) at intersection of temperature of air leaving cooling coils and straight line from sensible heat ratio *(SHR)* through point *A.*

3. Locate point *C* (condition of outdoor air).

4. Identify point *D* (return air mixture) at $\frac{1}{5}$ distance from point *A* along straight line *A-C* as make-up air is 20% in this example problem.

5. Find refrigeration load by formula, *H.G.*= 4.5 $Q\Delta h;$ where *Q* = total supply airflow volume in cfm and Δh = enthalpy in Btu/lb.

H.G. = 4.5 X 1160 X (31.5-22.7) = 46,000 Btuh or $\simeq$ ⟨4-tons⟩

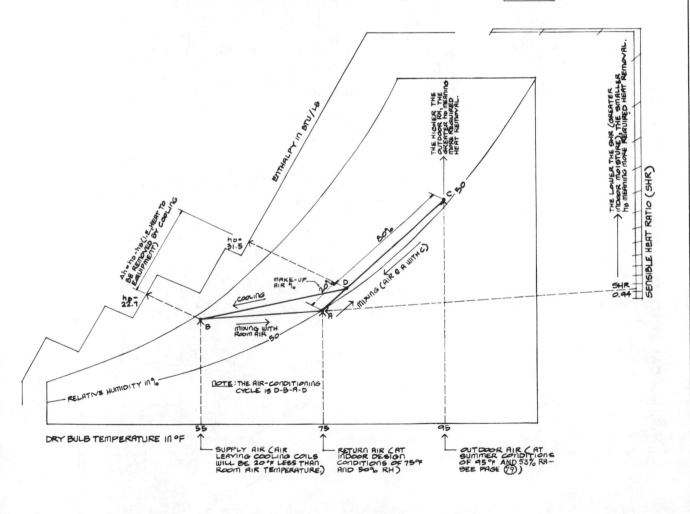

MECHANICAL SYSTEMS—PSYCHROMETRIC CHART

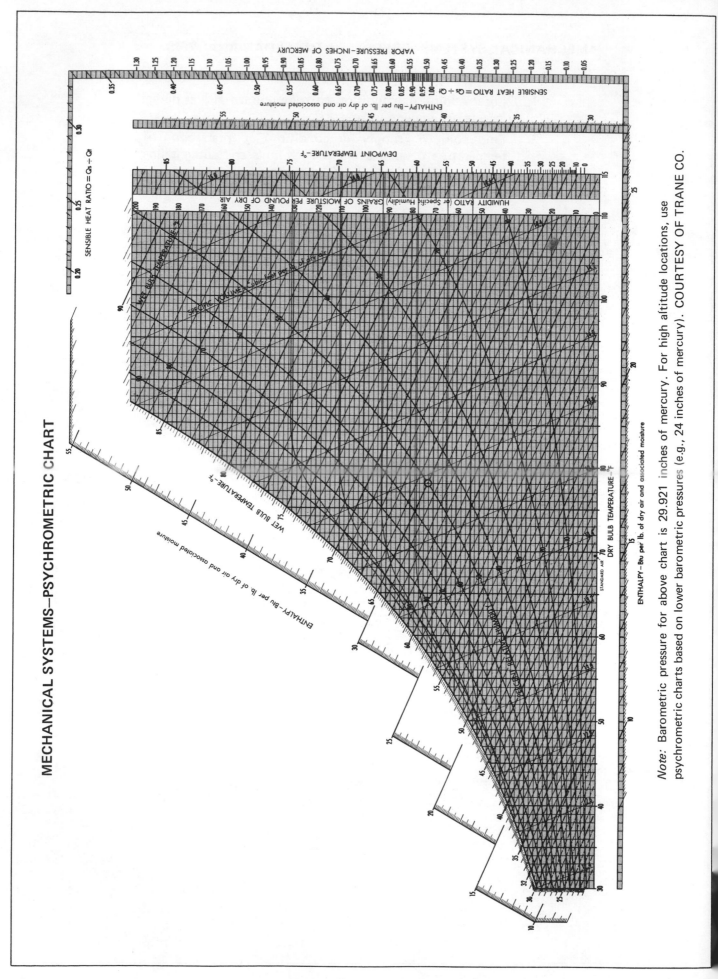

Note: Barometric pressure for above chart is 29.921 inches of mercury. For high altitude locations, use psychrometric charts based on lower barometric pressures (e.g., 24 inches of mercury). COURTESY OF TRANE CO.

MECHANICAL SYSTEMS—EFFECT OF BUILDING ORIENTATION ON COOLING LOADS

The table shows typical increases in cooling load for office buildings compared to the basic square shape shown below. The buildings, located at 40° north latitude, have equal floor areas. Note that heating loads for the square and diamond shapes will be about 10% less than the loads for all orientations of the rectangular shape.

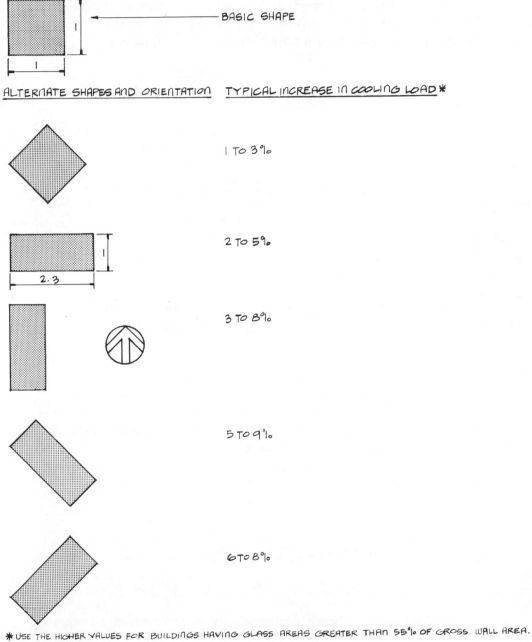

ALTERNATE SHAPES AND ORIENTATION   TYPICAL INCREASE IN COOLING LOAD *

BASIC SHAPE

1 TO 3%

2 TO 5%

3 TO 8%

5 TO 9%

6 TO 8%

* USE THE HIGHER VALUES FOR BUILDINGS HAVING GLASS AREAS GREATER THAN 55% OF GROSS WALL AREA.

REFERENCE

Sun, T. and K.S. Park, "Mechanical/Electrical." *Architectural Graphic Standards.* New York: John Wiley & Sons, Inc., 1971. Chaps. 15-16.

MECHANICAL SYSTEMS—EXHAUST-OUTDOOR AIR INTAKE BALANCE

Air is removed from buildings by exhaust hoods in kitchen areas, exhaust fans in toilet rooms, etc. In addition, conditioned air can be lost by exfiltration through openings around doors and windows. Outdoor air entering the conditioned space should equal the air lost through exhaust and exfiltration to provide positive building pressure. Systems that do *not* provide sufficient outdoor replacement (or "make-up") air will pull unfiltered and unconditioned air through every crack or opening and prevent exhaust fans from operating properly. Some areas, however, should have a slightly negative pressure (e.g., bathrooms, kitchens, etc.).

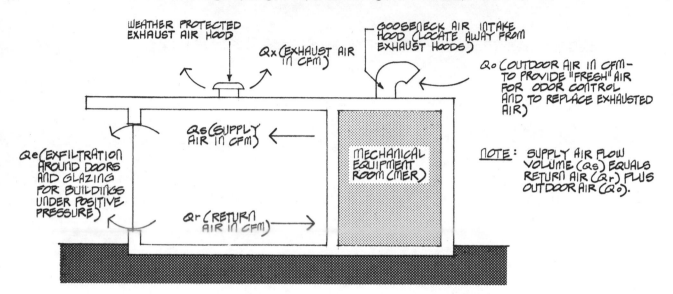

Example: A small luncheonette/cafeteria requires a supply airflow volume (Q_s) for cooling of 5,000 cfm. Exhaust hood above the short-order range removes 180 cfm, exhaust fans in the restrooms, 100 cfm. Assume an exfiltration rate of 125 cfm for all door and window cracks. What will be the return airflow volume? If make-up air is 10%, how much additional air could be exhausted?

Required outdoor air will be:

$$Q_o = 0.10\, Q_s = 0.10 \times 5{,}000 = \boxed{500 \text{ cfm}}$$

Find return air from formula, $Q_s = Q_r + Q_o$:

$$Q_r = Q_s - Q_o = 5{,}000 - 500 = \boxed{4{,}500 \text{ cfm}}$$

To provide positive pressure, additional exhausted air should not exceed:

$$Q = Q_o - (Q_x + Q_e) = 500 - (180 + 100 + 125) = \boxed{95 \text{ cfm}}$$

Note: See page 154 for a discussion of thermal recovery systems using heat wheels between the exhaust and supply air streams.

MECHANICAL SYSTEMS—HORIZONTAL AIR DISTRIBUTION EXAMPLES

Coordination of the air distribution paths with other services and building structure is essential to conserve space. Example integrated horizontal air duct networks for various ceiling systems are shown below.

Flat Panels

The floor-ceiling plenum space can be reduced by running air ducts between large lighting fixtures. In the example at the left, the required vertical clearance will be duct depth (d) plus about 12 in. for structure dimensional tolerance, duct insulation, duct reinforcing angles, and lighting fixtures.

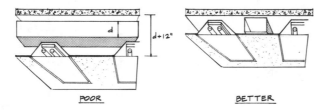

Baffles

Lighting baffle or louver screens can be used to support linear air diffusers. The baffles can also be used as a ceiling track to receive moveable partitions.

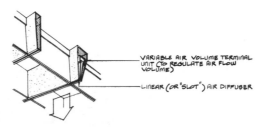

Coffers

Structural elements, such as coffered floor-ceilings, can be designed to include air paths. Vertical air distribution can be integrated with the column or pilaster design. Note that the formwork design is complex and should be fabricated with care. In-duct airflow control devices may require modification to be accessible for adjustment and maintenance.

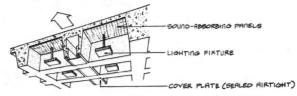

Note: Avoid air leaks by carefully sealing all cracks and openings (e.g., at the vertical distribution connections).

REFERENCE

Fischer, R. E. and F. J. Walsh, "Air Conditioning: A New Interpretation." *Architectural Record.* McGraw-Hill, Inc., 1970.

MECHANICAL SYSTEMS—EXPOSED DUCTWORK EXAMPLES

Air distribution ducts can be exposed within occupied spaces as shown
by the example layout drawings. Note that the plan view on the right
is a linear duct layout that can be integrated with structural framing
members and expressive lighting systems.

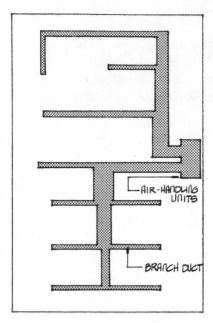

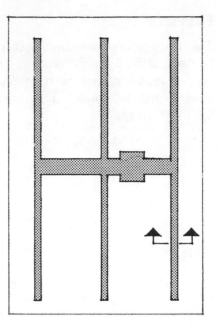

DISORGANIZED

BRANCH DUCT CROSS-SECTION
AREAS ARE REDUCED AS LESS
AIR IS CARRED BY SUCCESSIVE
BRANCHES.

ORGANIZED

SQUARE DUCTS, HAVING CONSTANT CROSS-
SECTION AREA, ARE COORDINATED WITH
THE LIGHTING SYSTEM, I.E., INDIRECT
FLUORESCENT LINE SOURCES. THIS LAYOUT
USES THE GREATER AMOUNT OF DUCT
MATERIAL, BUT WILL HAVE LOWER AIR
VELOCITIES DUE TO THE LARGER OVERALL
CROSS-SECTIONS.

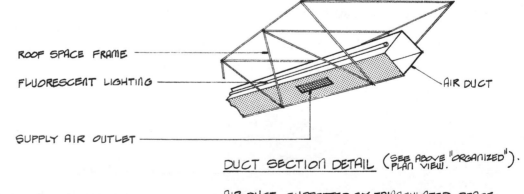

ROOF SPACE FRAME

FLUORESCENT LIGHTING

SUPPLY AIR OUTLET

AIR DUCT

DUCT SECTION DETAIL (SEE ABOVE "ORGANIZED" PLAN VIEW.)

AIR DUCT, SUPPORTED BY TRIANGULATED SPACE
FRAME ROOF, IN TURN SUPPORTS LIGHTING SYSTEM.

REFERENCE

Fischer, R. E., *Architectural Engineering-Environmental Control.*   New York:
McGraw-Hill, Inc., 1965, p. 28-32.

MECHANICAL SYSTEMS—COMMERCIALLY AVAILABLE AIR OUTLETS

Some of the many commercially available air outlet devices are shown below. Ceiling outlets, for example, come in various round, square, and rectangular shapes. Data for outlet airflow capacity (cfm), air velocity (fpm), throw distance (ft), pressure drop (''w.g.) and noise spectrum generated should be obtained from the manufacturer. Be careful to determine the specific noise ratings furnished and how the results were obtained. (See recommended air velocity table on following page.)

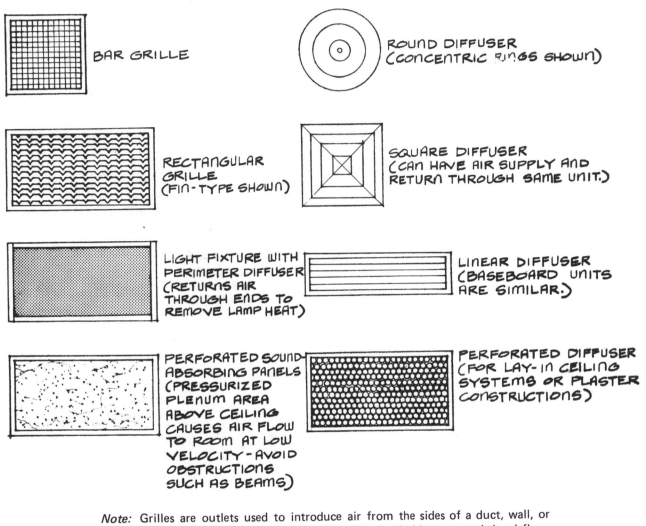

BAR GRILLE

ROUND DIFFUSER (CONCENTRIC RINGS SHOWN)

RECTANGULAR GRILLE (FIN-TYPE SHOWN)

SQUARE DIFFUSER (CAN HAVE AIR SUPPLY AND RETURN THROUGH SAME UNIT.)

LIGHT FIXTURE WITH PERIMETER DIFFUSER (RETURNS AIR THROUGH ENDS TO REMOVE LAMP HEAT)

LINEAR DIFFUSER (BASEBOARD UNITS ARE SIMILAR.)

PERFORATED SOUND-ABSORBING PANELS (PRESSURIZED PLENUM AREA ABOVE CEILING CAUSES AIR FLOW TO ROOM AT LOW VELOCITY - AVOID OBSTRUCTIONS SUCH AS BEAMS)

PERFORATED DIFFUSER (FOR LAY-IN CEILING SYSTEMS OR PLASTER CONSTRUCTIONS)

Note: Grilles are outlets used to introduce air from the sides of a duct, wall, or floor. A register is basically a grille with a damper behind it to control the airflow. Grilles and registers are available in various materials and types as shown by the examples above. Ceiling registers are generally called "diffusers" because they discharge air in a spreading pattern. Diffusers are generally required where room air change exceeds about 30 changes per hour.

SUGGESTED AIR VELOCITIES AT ROOM REGISTERS

Type of Space	Supply Register (in fpm)	Return Grille (in fpm)
Bedrooms, sleeping quarters, hospitals, residences, apartments, hotels, motels, auditoriums, theaters, radio/TV studios, music practice rooms, large meeting rooms, audio-visual facilities, large conference rooms, executive offices, churches, courtrooms, chapels, etc.	300 to 425	360 to 510
Private or semiprivate offices, small conference rooms, classrooms, reading rooms, libraries, etc.	425 to 500	510 to 600
Large offices, reception areas, retail shops and stores, cafeterias, restaurants, gymnasiums, etc.	500 to 575	600 to 690
Lobbies, corridors, laboratory work spaces, drafting and engineering rooms, general secretarial areas, maintenance shops (such as for electrical equipment), etc.	575 to 650	690 to 780
Kitchens, laundries, school and industrial shops, garages, machinery spaces, computer equipment rooms, etc.	650 to 725	780 to 870

The table above shows approximate air velocities in feet per minute (fpm) required to provide acceptable acoustical background noise levels at room registers. The table is intended for use with duct systems having internal glass-fiber linings and at least $\frac{1}{2}$ in. wide register slot openings. Note that the sound levels generated at room registers vary in direct proportion to the air velocity and therefore can be controlled by selecting registers of proper sizes. These velocities should be used for design purposes only when specific acoustical data for manufacturer's equipment are not available.

MECHANICAL SYSTEMS—AIR SUPPLY OUTLETS

The throw distance (*T*) for an air supply outlet depends on the velocity of the air and the size and shape of the outlet openings. Generally, select a throw distance of three-fourths the distance to the opposite wall. The number of outlets required is determined by the room air-flow volume (*Q*) and the outlet's air distribution pattern. Consult manufacturer's catalogs for specific outlet data including type of throw. If straight throw, the outlet spacing (*S*) should be about *T*/3; if fan-shaped or widespread, about *T*. Sometimes throw is referred to as "blow."

Outlet Placement and Dimensions (Wall Registers)

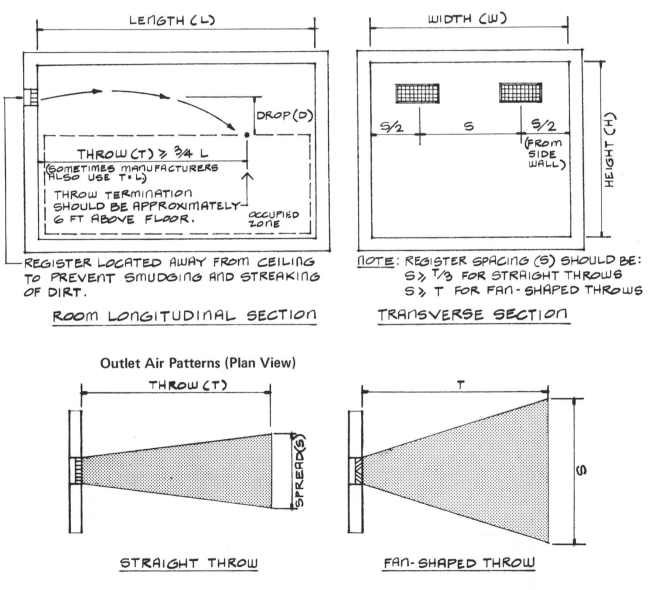

REGISTER LOCATED AWAY FROM CEILING TO PREVENT SMUDGING AND STREAKING OF DIRT.

ROOM LONGITUDINAL SECTION

NOTE: REGISTER SPACING (S) SHOULD BE:
S ≥ T/3 FOR STRAIGHT THROWS
S ≥ T FOR FAN-SHAPED THROWS

TRANSVERSE SECTION

Outlet Air Patterns (Plan View)

STRAIGHT THROW

FAN-SHAPED THROW

Note: Ceiling diffusers should be distributed so each unit will supply air to a specific area. Diffuser spacing (*S*) should about equal room height (*H*) for fan-shaped throw, about *H*/2 for straight throw. Check manufacturer's data for diffuser coverage pattern (i.e., spread radius, throw to each side, etc.) at design airflow conditions. Be careful to measure *H* as the distance from the floor to the lowest ceiling beam or obstruction.

MECHANICAL SYSTEMS—ROOM AIR DISTRIBUTION
EXAMPLE APPLICATIONS

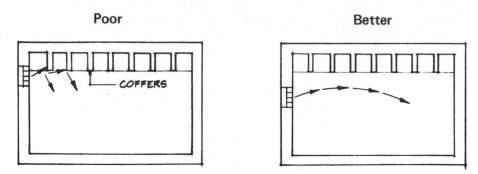

Poor   Better

Coffers

Ceiling coffers as shown above can affect the airflow from wall registers. To provide good air distribution, locate registers away from obstructions (or block off the upper portion of poorly located outlets if new air velocity will not exceed preferred range for noise control).

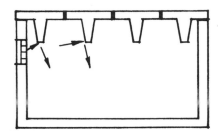 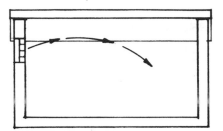

Ceiling Beams

Ceiling beams, large lighting fixtures or coves, etc. should not be positioned where they will interrupt airflow from wall registers as shown on the above figure at the left. Instead, locate wall outlets to throw air along the long axis of large linear elements.

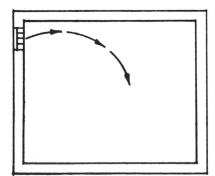 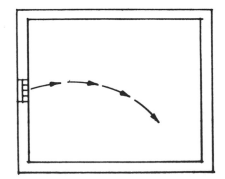

High Ceiling

In spaces with extremely high ceilings (e.g., banks, lobbies, gymnasiums) locate registers part way up walls to help keep some of the ceiling heat gain out of the occupied zone and to improve the air distribution pattern. Note that the occupied zone is usually defined as the room volume which extends to a height of about 6 ft and to within 6 in. of the walls, columns, and partitions.

RETURN AIR SYSTEMS

The design of most return air systems is not as detailed as supply duct systems since return air can be allowed to "float" to grille locations for reconditioning by the mechanical equipment. The return air grille location will not influence room air motion in spaces where supply air outlets are properly located. This situation is analogous to blowing out a lighted match held in your hand. Now try to extinguish another match under identical conditions this time by inhaling—it can not be done!

Often plenum spaces (above suspended ceilings or under floors) are used to return air as it can be exhausted into the mechanical equipment room by fans. In the cool climate region, consider using floor returns unless supplementary perimeter heat is available for exterior zones. It is *not* good practice to use corridors as return air plenums if door thresholds are undercut or louvers are placed between occupied spaces and the corridor. These direct openings will provide poor acoustical privacy. Where air is returned to a corridor, use separate internally lined duct branches or sound attenuating air-vent devices. Vent mufflers, for example, are commercially available that can be installed in walls, on doors, or in ceilings to prevent sound transmission.

Return air ducts can be sized using the design procedures presented earlier in this section. Be sure to account for the air exhausted from buildings and exfiltration (or leakage) through cracks around *all* doors and windows when establishing the return airflow volumes. Some laboratories and parts of hospitals (e.g., operating rooms, research areas) should *not* use recirculated air. To avoid cross-contamination, 100% outdoor air should be provided to these spaces. Recommended air velocities for duct systems are given on page 120. For data on return air grilles and registers, consult manufacturer's catalogs. Note that lengthy return air systems may require return air fans.

MECHANICAL SYSTEMS—AUTOMATIC CONTROLS

Automatic controls, integrated with mechanical systems, are used to obtain desired room temperatures, relative humidity, etc. The conditioned space may be controlled as a single area or be divided into two or more zones. Shown below are alternate room heating/cooling thermostat locations for a single zone dwelling unit. In commercial buildings, however, the best control arrangement for comfort *always* is individual control for each room.

Symbols:

Ⓣ Thermostat (temperature control device that operates fan switches, flow valves, dampers, etc.) location

⬡ Poor thermostat location

Ⓗ Humidistat (humidity control device that operates spray pumps, exhaust fan switches, etc.) location

Examples of Poor Thermostat Placement:

① On outside wall

② Opposite exterior glazing

③ Near kitchen range

④ Near lighting fixture, on outside wall

Checklist for Effective Use of Thermostats

Place thermostats on indoor wall locations about 4 to 6 ft above floor for protection from accidental impacts and ease in operation. Conditions at the thermostat should be representative of the area being controlled, e.g., near return air grille locations.

Do *not* place thermostats where they will be in the primary supply air stream, near exterior glazing (to avoid direct solar radiation), or near large lighting fixtures.

Avoid locations on shaft walls that enclose water pipes and walls near high temperature sources such as kitchen ranges, hair dryers, etc.

REFERENCE

Haines, Roger W., *Control Systems for HVAC.* New York: Van Nostrand Reinhold Co., 1971.

AIR CLEANING DEVICES

Dry Filters

Dry filters are usually manufactured from glass-fiber, wire screen, or cellulose. Filters are available that can be cleaned or disposed and replaced when they become dirty. Dry filters must be kept clean during use to allow air-handling units to operate at their design conditions because blocked filters can greatly reduce the supply air delivered to occupied spaces. High efficiency particulate air (abbreviated HEPA) filters, constructed from glass- and asbestos-fibers, were developed for the Atomic Energy Commission during the 1940's. Pressure drops through HEPA filters are usually high. High efficiency dry filters will remove tobacco smoke but *not* the irritating gases and vapors. Consequently, activated charcoal is often used for odor removal since gases and vapors can be adsorbed on the charcoal.

Spray Washers

Since most odors are due to a vapor of some compound mixed with air, they will dissolve in water. Spray washers introduce a mist of water into the supply air stream to dissolve the vapors. The water used in the washers must be changed frequently because it can become saturated with soluble vapors. Greasy particles, soot, and tobacco smoke are not water soluble and, therefore, will usually pass through spray washers.

Electric Precipitators

Electric precipitators consist of metal plates that give a positive charge to passing particles in the air stream. These positive charges are attracted by electrostatic forces to negatively charged plates which then can be manually cleaned with a solvent. Frequent cleaning of the collection plates is necessary.

Space-Charge Neutralization

Air-borne particles that have a positive charge will be attracted to grounded surfaces (e.g., walls, ceilings, furnishings) by electrostatic forces. Some common sources of positive charged particles are cigarette and cigar smoke, fluorescent lighting fixtures, air-conditioning equipment, and office machines. Since the buildup of charged particles causes significant odor problems and staining of room furnishings (even when conventional cleaning devices are used), it is important that the charges be effectively neutralized. Space-charge neutralization techniques are available that will dilute positive charges on air-borne particles by passing the room supply air through an electrode combination consisting of a high-voltage, high frequency source of capacitance. The neutralized particles then may be effectively removed by conventional filters.

MECHANICAL SYSTEMS—CLEAN SPACES

Clean spaces are needed in hospital operating rooms, computer equipment rooms, sophisticated manufacturing and assembly plants to control particulate contamination. Clean spaces may be entire rooms or even individual work stations within rooms. Examples of clean space designs for both conventional and laminar airflow patterns are given below. In the laminar system, air is supplied in parallel flow lines at uniform velocity.

Conventional Flow Systems

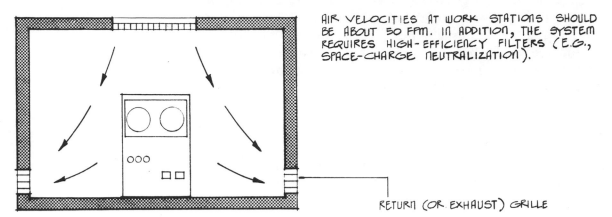

AIR VELOCITIES AT WORK STATIONS SHOULD BE ABOUT 50 FPM. IN ADDITION, THE SYSTEM REQUIRES HIGH-EFFICIENCY FILTERS (E.G., SPACE-CHARGE NEUTRALIZATION).

RETURN (OR EXHAUST) GRILLE

Laminar Flow Systems

Air velocities at work stations should be in the range of 70 to 120 fpm to prevent settling of particles (note that 120 fpm is about $1\frac{1}{2}$ mph). In laminar systems, air supply comes directly through high efficiency particulate air (HEPA) filters, *not* through ducts that can allow the supply air to become contaminated before it reaches the air outlets.

SUPPLY CEILING OUTLET (ENTIRE CEILING NORMALLY CONSISTS OF HEPA FILTERS.)

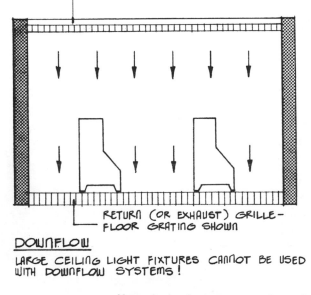

RETURN (OR EXHAUST) GRILLE—FLOOR GRATING SHOWN

<u>DOWNFLOW</u>

LARGE CEILING LIGHT FIXTURES CANNOT BE USED WITH DOWNFLOW SYSTEMS!

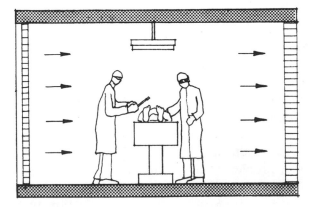

<u>CROSSFLOW</u>

OFTEN PARTICULATE-GENERATING ACTIVITY CAN BE PUT OFF TO ONE SIDE OR DOWN-STREAM FROM THE WORK STATION.

Note: In laminar systems, thermal comfort of occupants is generally independent of air stream direction.

MECHANICAL SYSTEMS—TOTAL ENERGY SYSTEMS

Shown below are the basic elements of a total energy system using a natural gas (or fuel oil) powered turbine to drive an electric generator. The heat produced by the turbine is reclaimed by a boiler which in turn is used for heating and cooling. The total energy plant usually requires 10 to 15% more space than a conventional system.

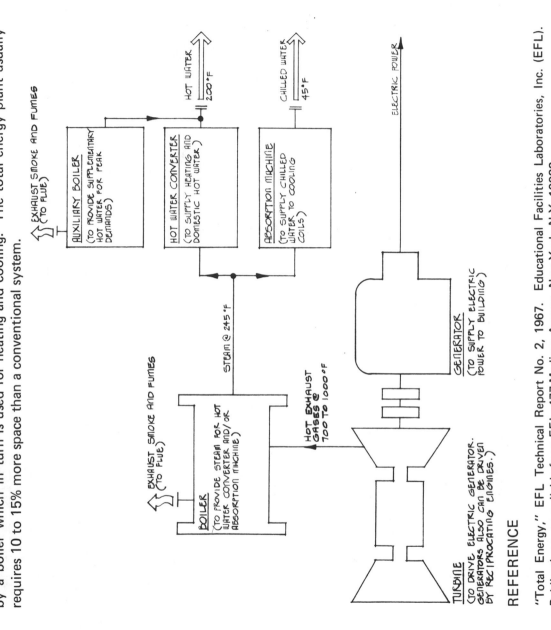

REFERENCE

"Total Energy," EFL Technical Report No. 2, 1967. Educational Facilities Laboratories, Inc. (EFL). Publications are available from EFL, 477 Madison Avenue, New York, N Y 10022.

MECHANICAL SYSTEMS—COOLING: ABSORPTION REFRIGERATION CYCLE

Cooling by absorption is shown by the diagram below. Chilled water is produced by using an absorbent salt solution (usually lithium bromide) and by controlling its properties with hot water and cooling tower water. An inexpensive source of hot water or steam is required to make the absorption cycle economically competitive with the compressive cycle (See page 97).

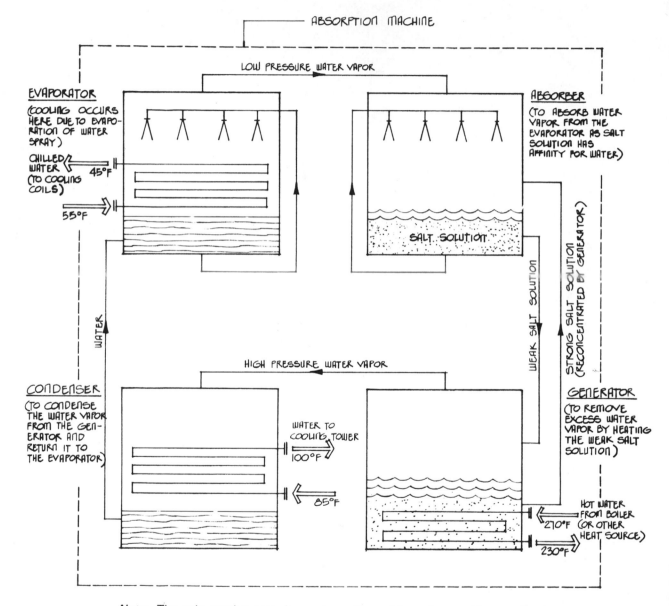

Note: The only moving parts in an absorption refrigeration machine are the circulating pumps. Nevertheless, it should be supported by resilient isolators (See Section 6).

MECHANICAL SYSTEMS—SOLAR ENERGY SYSTEMS

Solar energy systems absorb energy from solar radiation for building heating. In the overheated season, a solar system can be used to supply hot water to an absorption refrigeration machine (or a heat pump with solar collector used as condenser) for building cooling. Be careful to place solar collectors where they will *not* be shaded for a considerable part of the day by nearby terrain or structures (See page 27).

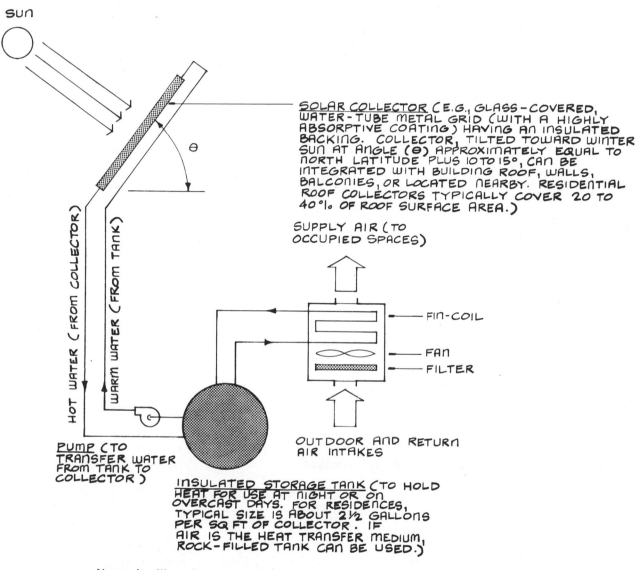

SUN

SOLAR COLLECTOR (E.G., GLASS-COVERED, WATER-TUBE METAL GRID (WITH A HIGHLY ABSORPTIVE COATING) HAVING AN INSULATED BACKING. COLLECTOR, TILTED TOWARD WINTER SUN AT ANGLE (θ) APPROXIMATELY EQUAL TO NORTH LATITUDE PLUS 10 TO 15°, CAN BE INTEGRATED WITH BUILDING ROOF, WALLS, BALCONIES, OR LOCATED NEARBY. RESIDENTIAL ROOF COLLECTORS TYPICALLY COVER 20 TO 40% OF ROOF SURFACE AREA.)

HOT WATER (FROM COLLECTOR)

WARM WATER (FROM TANK)

SUPPLY AIR (TO OCCUPIED SPACES)

FIN-COIL

FAN

FILTER

OUTDOOR AND RETURN AIR INTAKES

PUMP (TO TRANSFER WATER FROM TANK TO COLLECTOR)

INSULATED STORAGE TANK (TO HOLD HEAT FOR USE AT NIGHT OR ON OVERCAST DAYS. FOR RESIDENCES, TYPICAL SIZE IS ABOUT 2½ GALLONS PER SQ FT OF COLLECTOR. IF AIR IS THE HEAT TRANSFER MEDIUM, ROCK-FILLED TANK CAN BE USED.)

Note: Auxiliary heat sources (e.g., electric resistance heaters) are designed to operate whenever inside air temperature drops and to provide full heating load if required.

REFERENCE

Yellott, J. I., "Utilization of Sun and Sky Radiation for Heating and Cooling of Buildings." *ASHRAE Journal,* December 1973.

MECHANICAL SYSTEMS—HEAT RECOVERY SYSTEM EXAMPLES

Heat Wheels

The wheel is packed with heat-absorbing material such as aluminum or stainless steel shavings. This material absorbs heat from the warm building exhaust air and, by rotation of the wheel, transfers this heat to the colder intake air. (In summer months, the wheel also can be used to cool incoming air.) Because a motor is required to turn the heat wheel, heat recovery is not without operational cost.

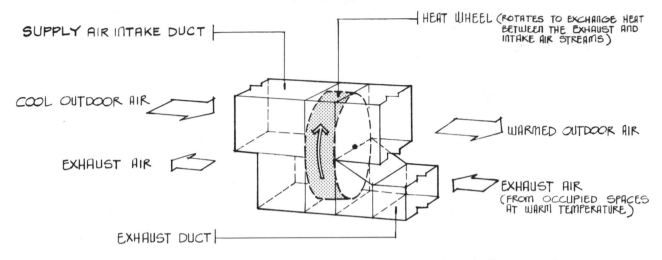

Note: Where exhaust and intake air ducts are *not* close together, the "run-around coil" system can be used to transfer heat energy. In this system, a fin-coil, located in the exhaust duct, heats (or cools) liquid that is circulated to a fin-coil in the intake duct.

Heat From Light Sources

In the examples shown below return air is drawn through the lighting fixtures to remove heat before it enters the occupied zone. The heated air in the ceiling plenum space can be exhausted or removed by recirculation to occupied zones needing heat. In addition, light sources are more efficient at lower temperatures.

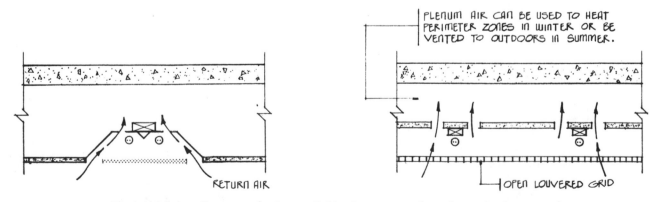

Note: Lighting fixtures are also available that remove heat from the lamps and ballasts by circulating water through built-in fixture tubing that can be connected to a cooling unit.

MECHANICAL SYSTEMS—AIR-CURTAINS

An air-curtain is a directional layer of high velocity air blown across an opening to retard the movement of heat, moisture, dirt, etc. into conditioned spaces. Air-curtains used for pedestrian openings in retail shops and stores are generally the downflow type (shown below) having air velocities in the range of 800 to 1,200 fpm (about 10 to 15 mph). Industrial applications can be either downflow or horizontal flow types, usually at much greater air velocities. Air-curtain systems may have either ducted or non-ducted returns.

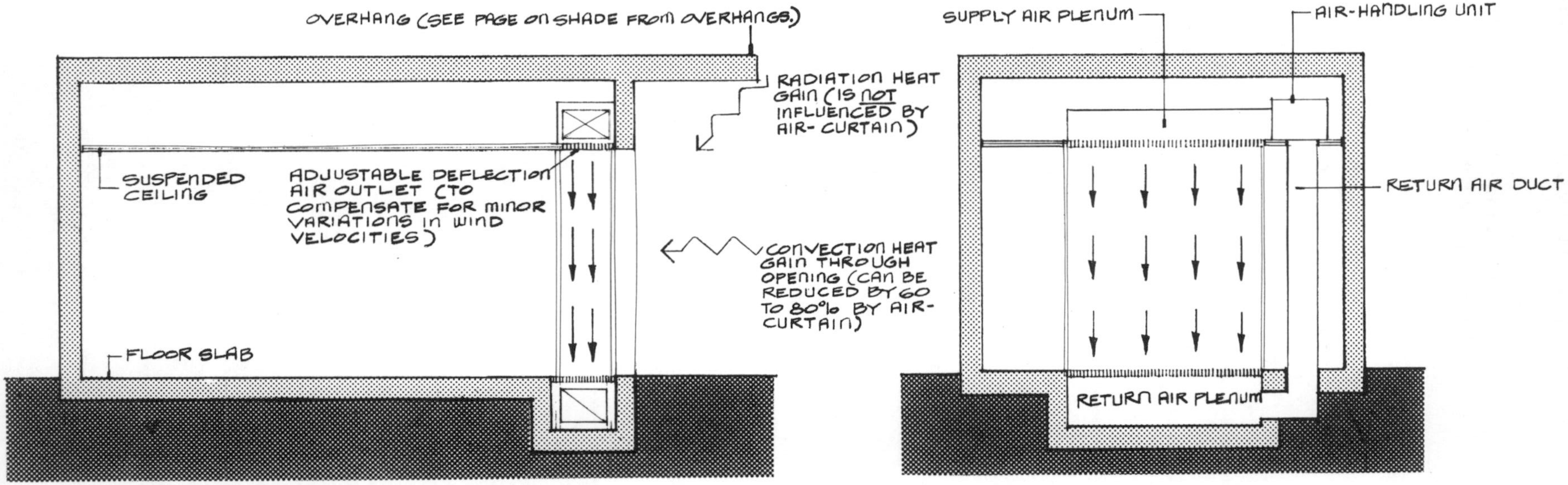

AIR-HANDLING UNIT

RETURN AIR DUCT

SUPPLY AIR PLENUM

RETURN AIR PLENUM

OVERHANG (SEE PAGE ON SHADE FROM OVERHANGS)

RADIATION HEAT GAIN (IS NOT INFLUENCED BY AIR-CURTAIN)

CONVECTION HEAT GAIN THROUGH OPENING (CAN BE REDUCED BY 60 TO 80% BY AIR-CURTAIN)

SUSPENDED CEILING

ADJUSTABLE DEFLECTION AIR OUTLET (TO COMPENSATE FOR MINOR VARIATIONS IN WIND VELOCITIES)

FLOOR SLAB

Note: In industrial applications where large areas must be open to conditioned spaces for loading, off-loading, etc., two sets of doors, with sufficient working space between, may be used to create a "thermal air-lock."

CHECKLIST FOR MECHANICAL SYSTEMS

When realistic provisions for mechanical space requirements are determined during the design stages of a project, only minor adjustments should be needed after the actual equipment is selected. The final equipment layout should be arranged to conserve space and provide clearance for equipment service and repair.

In general, locate mechanical equipment spaces centrally with access to the outdoors. For noise control within buildings, mechanical spaces should be located, both horizontally and vertically, as far away as possible from areas requiring quiet acoustical backgrounds (e.g., bedrooms, conference rooms, etc.).

Air velocities in ducts should be low to conserve energy and to prevent noise. Consequently, in ductwork for high velocity all-air and air-water systems provide internal glass-fiber linings and sound attenuating mufflers.

Generally, ducts should be insulated with internal glass-fiber linings for condensation control as well as noise control. Where ducts pass through plenum and attic spaces that are not conditioned, they must be lined. Avoid crisscrossing of air ducts. Provide smooth turns in duct systems.

Locate supply air outlets away from obstructions such as ceiling coffers, beams, and lighting fixtures. Obstructions can cause uneven, ineffective distribution of air within conditioned spaces.

Large requirements of outdoor air will greatly increase building conditioning loads. Air quality can be achieved at lower airflow volumes by using odor-adsorbing devices such as activated charcoal filters.

Heat recovery and conservation techniques are available that can help to conserve energy resources when properly applied. For example, use automatic controls to operate exhaust hoods only when required.

MECHANICAL SYSTEM NOISE AND VIBRATIONS

MECHANICAL SYSTEM NOISE AND VIBRATIONS—VIBRATING EQUIPMENT

The vibration produced in buildings by mechanical equipment can be felt and heard by building occupants. Vibrations can travel through solid members such as columns, beams, etc. that may reradiate sound at great distances from the original source. It is therefore important that the vibrating equipment be properly isolated from the building structure with resilient mounts. If possible, vibrating equipment should be positioned away from the center of floor spans and relocated near columns or load bearing walls where they will have better structural support.

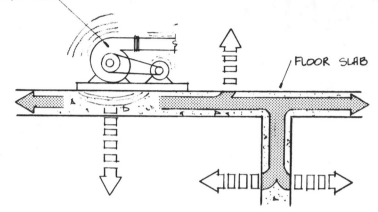

VIBRATING FAN

FLOOR SLAB

• <u>POOR</u>

FAN BOLTED TO FLOOR TRANSMITS VIBRATIONS DIRECTLY INTO STRUCTURE.

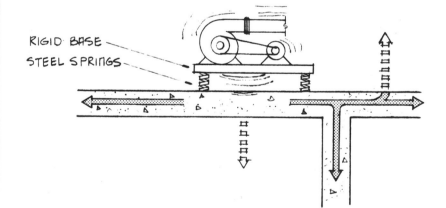

RIGID BASE

STEEL SPRINGS

• <u>BETTER</u>

FAN IS SUPPORTED BY RESILIENT MOUNTS AND RELOCATED CLOSE TO STRUCTURAL COLUMN. THE FAN CONTINUES TO VIBRATE, BUT THE "STRUCTURE-BORNE" SOUND IS REDUCED CONSIDERABLY.

MECHANICAL SYSTEM NOISE AND VIBRATIONS—BASIC VIBRATION THEORY

The principle of vibration isolation involves supporting the vibrating equipment by resilient materials such as: ribbed neoprene pads, pre-compressed glass-fiber pads, steel springs, etc. The goal is to choose a proper resilient material that, when loaded, will provide a *natural frequency* that is three or more times lower than the *driving frequency* of the equipment. Driving frequency is an operational characteristic of the equipment that should be obtained from the manufacturer. Natural frequency (f_n) in cycles per second (or hertz) can be calculated for steel springs since it varies with deflection as follows:

$$f_n = 3.13\sqrt{1/Y}$$

where f_n = natural frequency in Hz (cps)

   Y = static deflection in inches

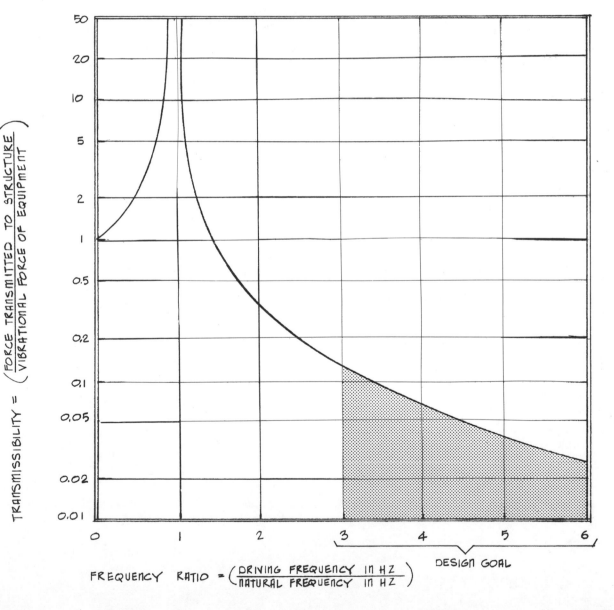

159

MECHANICAL SYSTEM NOISE AND VIBRATIONS—NATURAL FREQUENCY AND DEFLECTION

Curves below show natural frequency (f_n) for free-standing steel springs, precompressed glass-fiber, rubber, and cork. The most resilient isolators are springs because they have the largest deflections. Static deflection is the distance an isolator will compress (or deflect) when weight is applied to it.

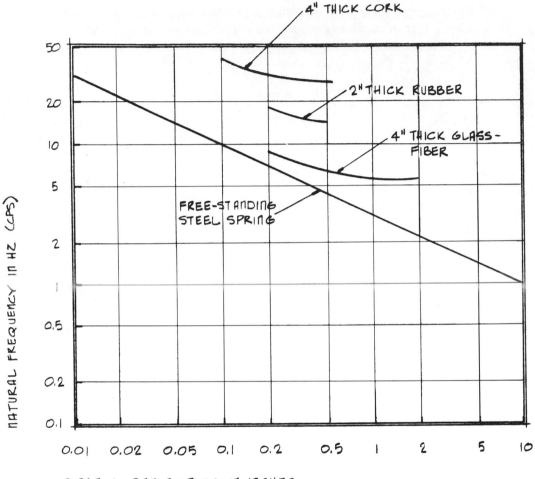

MECHANICAL SYSTEM NOISE AND VIBRATIONS—VIBRATION ISOLATION DESIGN GRAPH

The graph below can be used to find the static deflection required for vibrating equipment isolators. For example, an exhaust air fan (for a gymnasium) operating at a driving frequency of 520 cpm will require resilient isolators having a static deflection of at least 1 in. (See dashed lines on graph.) The graph is based on the assumption that the resilient materials will be in turn supported by a rigid base. Non-rigid, lightweight base supports such as above-grade mechanical equipment spaces (especially lightweight steel or wood-frame floor systems) require special consideration.

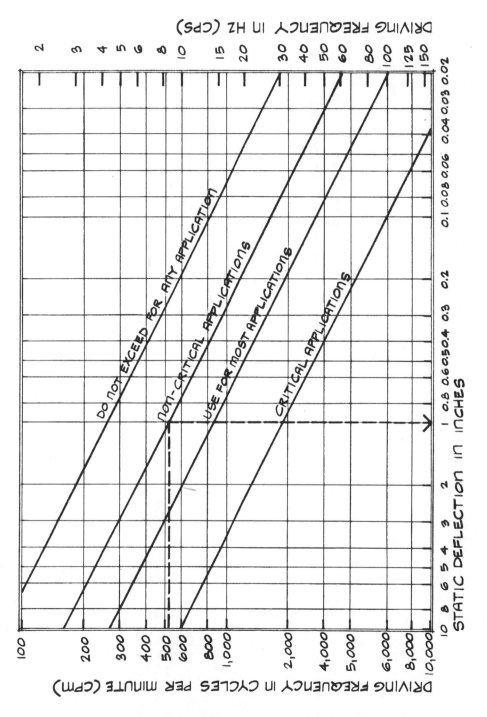

MECHANICAL SYSTEM NOISE AND VIBRATIONS—MECHANICAL DUCT SYSTEMS

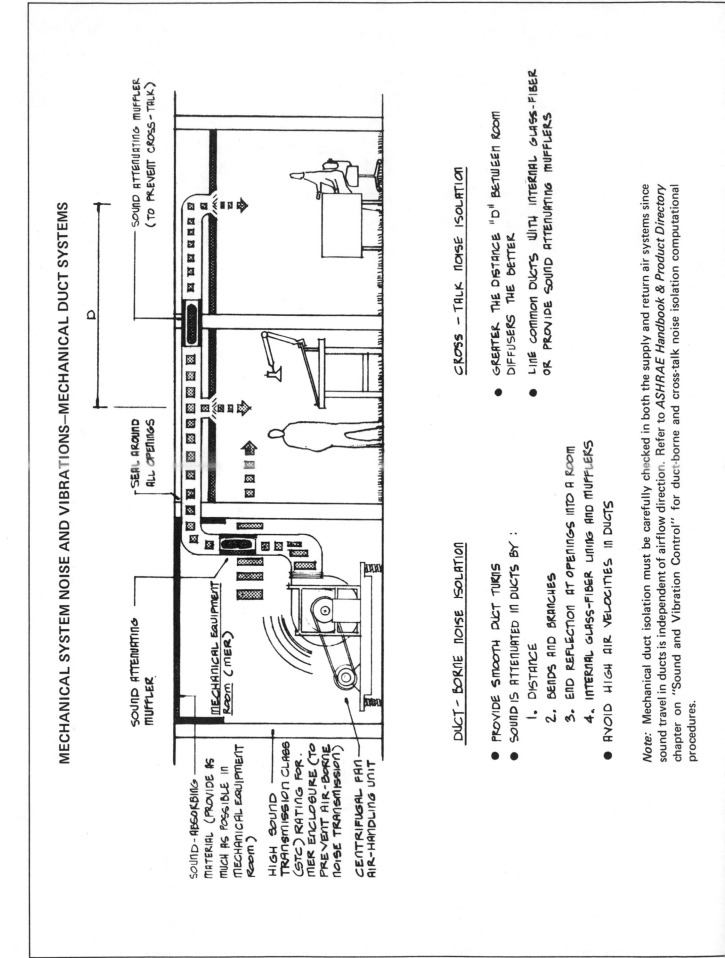

SOUND ATTENUATING MUFFLER (TO PREVENT CROSS-TALK)

SEAL AROUND ALL OPENINGS

SOUND ATTENUATING MUFFLER

MECHANICAL EQUIPMENT ROOM (MER)

SOUND-ABSORBING MATERIAL (PROVIDE AS MUCH AS POSSIBLE IN MECHANICAL EQUIPMENT ROOM)

HIGH SOUND TRANSMISSION CLASS (STC) RATING FOR MER ENCLOSURE (TO PREVENT AIR-BORNE NOISE TRANSMISSION)

CENTRIFUGAL FAN AIR-HANDLING UNIT

DUCT-BORNE NOISE ISOLATION

- PROVIDE SMOOTH DUCT TURNS
- SOUND IS ATTENUATED IN DUCTS BY:
  1. DISTANCE
  2. BENDS AND BRANCHES
  3. END REFLECTION AT OPENINGS INTO A ROOM
  4. INTERNAL GLASS-FIBER LINING AND MUFFLERS
- AVOID HIGH AIR VELOCITIES IN DUCTS

CROSS-TALK NOISE ISOLATION

- GREATER THE DISTANCE "D" BETWEEN ROOM DIFFUSERS THE BETTER
- LINE COMMON DUCTS WITH INTERNAL GLASS-FIBER OR PROVIDE SOUND ATTENUATING MUFFLERS

Note: Mechanical duct isolation must be carefully checked in both the supply and return air systems since sound travel in ducts is independent of airflow direction. Refer to *ASHRAE Handbook & Product Directory* chapter on "Sound and Vibration Control" for duct-borne and cross-talk noise isolation computational procedures.

MECHANICAL SYSTEM NOISE AND VIBRATIONS—MECHANICAL EQUIPMENT ROOM TREATMENT

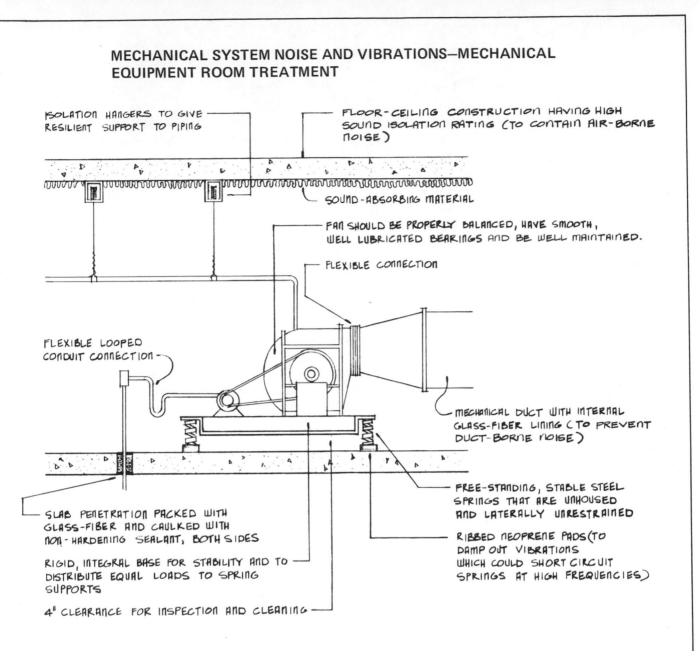

ISOLATION HANGERS TO GIVE RESILIENT SUPPORT TO PIPING

FLOOR-CEILING CONSTRUCTION HAVING HIGH SOUND ISOLATION RATING (TO CONTAIN AIR-BORNE NOISE)

SOUND-ABSORBING MATERIAL

FAN SHOULD BE PROPERLY BALANCED, HAVE SMOOTH, WELL LUBRICATED BEARINGS AND BE WELL MAINTAINED.

FLEXIBLE CONNECTION

FLEXIBLE LOOPED CONDUIT CONNECTION

MECHANICAL DUCT WITH INTERNAL GLASS-FIBER LINING (TO PREVENT DUCT-BORNE NOISE)

FREE-STANDING, STABLE STEEL SPRINGS THAT ARE UNHOUSED AND LATERALLY UNRESTRAINED

SLAB PENETRATION PACKED WITH GLASS-FIBER AND CAULKED WITH NON-HARDENING SEALANT, BOTH SIDES

RIBBED NEOPRENE PADS (TO DAMP OUT VIBRATIONS WHICH COULD SHORT CIRCUIT SPRINGS AT HIGH FREQUENCIES)

RIGID, INTEGRAL BASE FOR STABILITY AND TO DISTRIBUTE EQUAL LOADS TO SPRING SUPPORTS

4" CLEARANCE FOR INSPECTION AND CLEANING

SOME VIBRATION ISOLATION GUIDELINES

Fans

Large centrifugal fans can be a source of considerable low frequency vibration. These fans and their motors should be mounted on a common, rigid base to avoid mis-alignment, which wears out fan belts and bearings. The rigid base is in turn placed on spring isolators. Sometimes the base is a concrete slab called an "inertia block." Inertia blocks for fans (or pumps) are sized at 1 to 2 times the equipment weight plus all ductwork, conduit, etc. to the first hangers. The inertia block evens out the load on the springs as well as providing a rigid base. Additionally, its mass reduces the amplitude of vibration, but not the forces, and contributes somewhat to the air-borne sound transmission loss through the floor slab below.

Refrigeration Compressors

Large, low-speed reciprocating compressors should be isolated by springs and inertia blocks. High-speed centrifugal compressors require less isolation and often can be isolated properly with several layers of ribbed neoprene.

Cooling Towers

Cooling tower vibration involves the low frequency vibration of propeller-type fans as well as the high frequency waterfall noise. Large-deflection steel springs and ribbed neoprene mounts are required. Locate ground level cooling towers a considerable distance away from buildings to prevent air-borne noise problems.

Transformers

Transformers should be supported with several layers of ribbed neoprene. All electrical connections should be of flexible braided cable and conduits should have flexible joints. Also, partial or demountable enclosures, lined with sound-absorbing materials, can be used to help isolate annoying humming noises.

REFERENCES

1. *ASHRAE Handbook & Product Directory: Systems,* American Society of Heating, Refrigerating and Air-Conditioning Engineers, Inc. (ASHRAE), 1973. Published periodically at 345 E. 47th Street, New York, N Y 10017.
2. Fischer, R. E., "Some Particular Problems of Noise Control," *Architectural Record*, September 1968.

TYPICAL MECHANICAL EQUIPMENT NOISE IN DECIBELS

For many practical problems, the typical mechanical equipment decibel (dB)* values at 3 ft away given below may be used for design purposes if proper consideration is given to situations that may exceed them. For example, peak values for especially large mechanical equipment may exceed the values in the table by 5 or more dB. Although the table values represent data from the upper end of noise level ranges for specific equipment types, data from field measurements or laboratory tests according to current ASHRAE standards are *always* preferred.

*The decibel (dB) is a dimensionless unit for expressing the ratio of sound energies on a logarithmic scale. The reference energy is usually taken at the threshold of hearing. Human perception of noise depends on frequency as well as energy level. Therefore, attenuation data used for designing noise control measures are stated in terms of dB at octave-band frequencies in hertz (Hz).

Typical Sound Pressure Level (dB)

Equipment	63 Hz	125 Hz	250 Hz	500 Hz	1000 Hz	2000 Hz	4000 Hz	8000 Hz
1. Absorption machine	91	86	86	86	83	80	77	72
2. Axial fan	98	99	99	98	97	95	91	87
3. Boiler	92	92	89	86	83	80	77	74
4. Centrifugal fan	86	95	89	90	87	82	76	77
5. Chiller, centrifugal	80	85	87	87	90	98	91	87
6. Compressor	86	84	86	87	86	84	80	75
7. Condenser	99	92	90	90	89	85	76	68
8. Cooling tower	102	102	97	94	90	88	84	79
9. Fan-coil unit	57	55	53	50	48	42	38	32
10. Induction unit	57	58	56	54	45	40	35	33
11. PTAC	64	64	65	56	53	48	44	37
12. Pump	75	80	82	87	86	80	77	75
13. Roof-top unit	95	93	89	85	80	75	69	66
14. Warm-air furnace	65	65	59	53	48	45	39	30

Source: *Noise from Construction Equipment and Operations, Building Equipment, and Home Appliances* (Washington, D C : U.S. Environmental Protection Agency, December 1971).

MECHANICAL SYSTEM NOISE AND VIBRATIONS—CROSS-TALK

Untreated mechanical ducts can act like speaking tubes to transmit unwanted sound from one room to another as shown below. The sound transmission loss through common ducts should equal the common wall transmission loss. Use duct linings or sound attenuating mufflers where required.

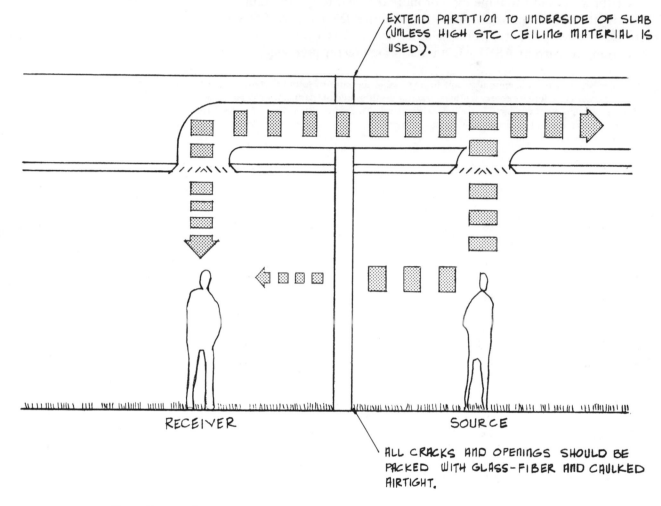

EXTEND PARTITION TO UNDERSIDE OF SLAB (UNLESS HIGH STC CEILING MATERIAL IS USED).

RECEIVER

SOURCE

ALL CRACKS AND OPENINGS SHOULD BE PACKED WITH GLASS-FIBER AND CAULKED AIRTIGHT.

Note: Ductwork should not be rigidly connected to partitions since they may transmit mechanical vibrations.

MECHANICAL SYSTEM NOISE AND VIBRATIONS—GLASS-FIBER DUCT LININGS

Sound attenuation from glass-fiber linings is often required in mechanical ducts to prevent transmission of equipment noise as well as cross-talk between rooms. The curves below show attenuation in decibels *per foot* of duct length for internal glass-fiber linings. The curves show 1 in. and 2 in. thick linings for a 16 in. deep mechanical duct. Actual data should be used whenever possible. For example, the attenuation shown below would be less for larger ducts with identical linings. (See note on page 165 for definition of decibel.)

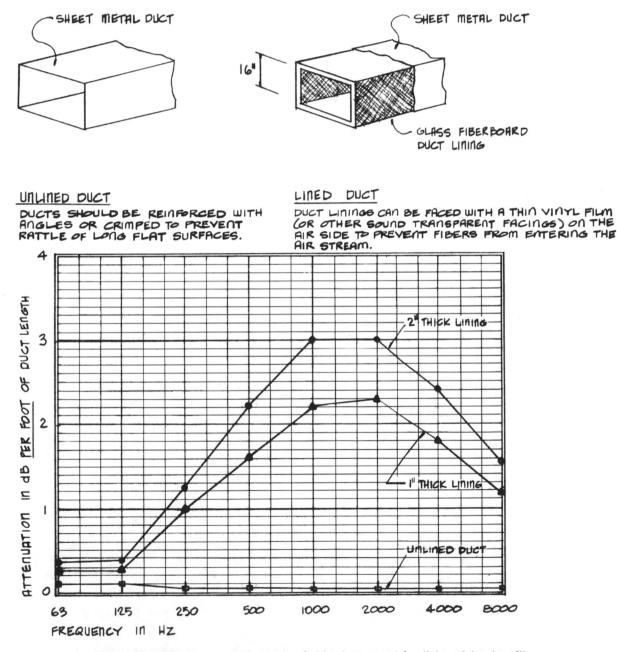

UNLINED DUCT
DUCTS SHOULD BE REINFORCED WITH ANGLES OR CRIMPED TO PREVENT RATTLE OF LONG FLAT SURFACES.

LINED DUCT
DUCT LININGS CAN BE FACED WITH A THIN VINYL FILM (OR OTHER SOUND TRANSPARENT FACINGS) ON THE AIR SIDE TO PREVENT FIBERS FROM ENTERING THE AIR STREAM.

Note: Since ducts are usually made of thin sheet metal (or lightweight glass fiber-board) sound can easily pass through the duct wall. Often an outer cover of plaster may be needed to improve isolation where ducts run through noisy spaces.

MECHANICAL SYSTEM NOISE AND VIBRATIONS—SOUND ATTENUATING MUFFLERS

Prefabricated sound attenuating mufflers are especially useful where high attenuation over a wide frequency range is required and the available length of duct is limited. They are normally available in both rectangular and round shapes. Typical sound attenuating data for a 3 ft long muffler is shown below. Refer to manufacturer's test data for anticipated performance at actual design conditions. Also, be sure to check manufacturer's test data for the noise generated by the muffler itself. Airflow through mufflers produces sound (from turbulence) that may be critical in applications where low acoustical background noise levels are required.

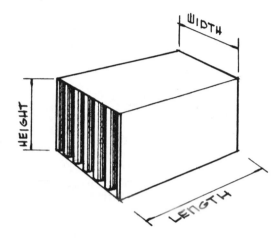
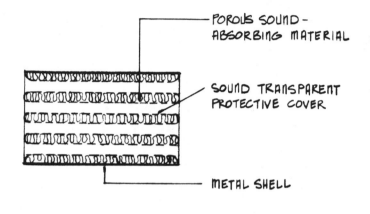

HORIZONTAL SECTION

RECTANGULAR MUFFLER

Sound attenuation in dB

Frequency (Hz):	125	250	500	1000	2000	4000	8000
Attenuation (dB): 3 ft long muffler	13	28	36	39	40	39	27

MECHANICAL SYSTEM NOISE AND VIBRATIONS—WALL CONSTRUCTIONS FOR MER's

The sound transmission class (STC) is a single-number rating of a construction's air-borne sound transmission performance evaluated over a standard frequency range. The higher the STC rating, the more efficient the construction for reducing sound transmission in the test frequency range. STC's for a construction should be obtained from field measurements or laboratory tests according to current ASTM standards. Shown below, in order of increasing STC's, are typical concrete block wall constructions for mechanical spaces. Note that the more massive a construction, the greater its resistance to air-borne sound.

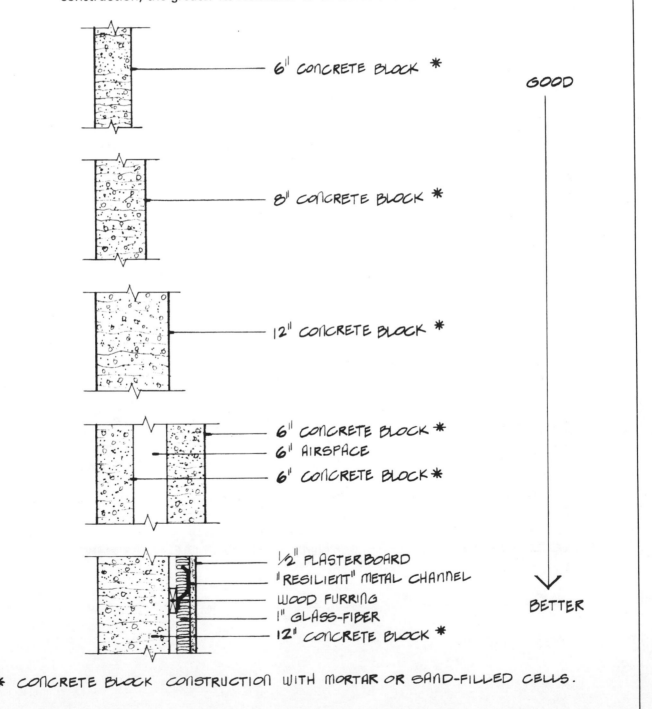

6" CONCRETE BLOCK *

8" CONCRETE BLOCK *

12" CONCRETE BLOCK *

6" CONCRETE BLOCK *
6" AIRSPACE
6" CONCRETE BLOCK *

½" PLASTERBOARD
"RESILIENT" METAL CHANNEL
WOOD FURRING
1" GLASS-FIBER
12" CONCRETE BLOCK *

GOOD

BETTER

* CONCRETE BLOCK CONSTRUCTION WITH MORTAR OR SAND-FILLED CELLS.

MECHANICAL SYSTEM NOISE AND VIBRATIONS—FLOOR-CEILING CONSTRUCTIONS FOR MER's

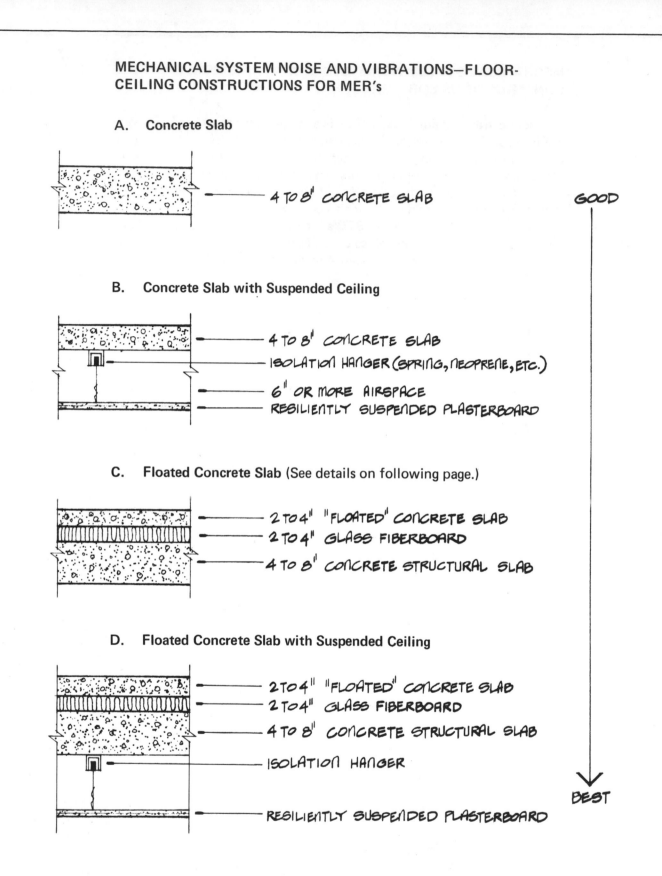

A. **Concrete Slab**

4 TO 8" CONCRETE SLAB

B. **Concrete Slab with Suspended Ceiling**

4 TO 8" CONCRETE SLAB
ISOLATION HANGER (SPRING, NEOPRENE, ETC.)
6" OR MORE AIRSPACE
RESILIENTLY SUSPENDED PLASTERBOARD

C. **Floated Concrete Slab** (See details on following page.)

2 TO 4" "FLOATED" CONCRETE SLAB
2 TO 4" GLASS FIBERBOARD
4 TO 8" CONCRETE STRUCTURAL SLAB

D. **Floated Concrete Slab with Suspended Ceiling**

2 TO 4" "FLOATED" CONCRETE SLAB
2 TO 4" GLASS FIBERBOARD
4 TO 8" CONCRETE STRUCTURAL SLAB
ISOLATION HANGER
RESILIENTLY SUSPENDED PLASTERBOARD

GOOD

BEST

MECHANICAL SYSTEM NOISE AND VIBRATIONS—FLOATED FLOOR DETAILS

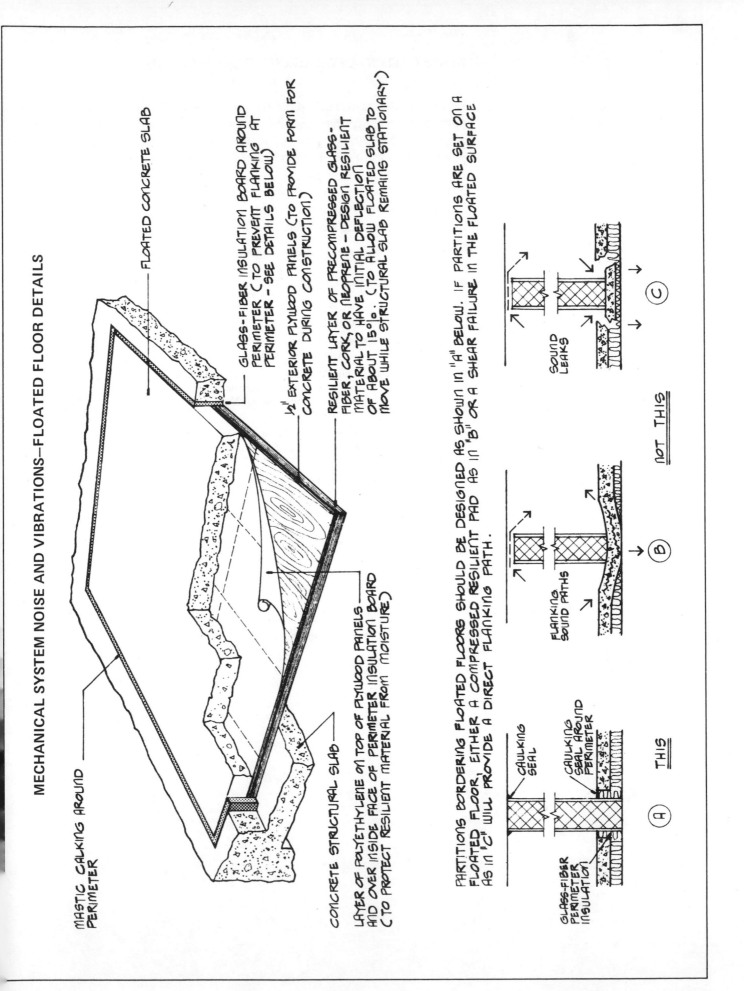

MASTIC CALKING AROUND PERIMETER

FLOATED CONCRETE SLAB

GLASS-FIBER INSULATION BOARD AROUND PERIMETER (TO PREVENT FLANKING AT PERIMETER - SEE DETAILS BELOW)

½" EXTERIOR PLYWOOD PANELS (TO PROVIDE FORM FOR CONCRETE DURING CONSTRUCTION)

RESILIENT LAYER OF PRECOMPRESSED GLASS-FIBER, CORK, OR NEOPRENE - DESIGN RESILIENT MATERIAL TO HAVE INITIAL DEFLECTION OF ABOUT 15%. (TO ALLOW FLOATED SLAB TO MOVE WHILE STRUCTURAL SLAB REMAINS STATIONARY)

CONCRETE STRUCTURAL SLAB

LAYER OF POLYETHYLENE ON TOP OF PLYWOOD PANELS AND OVER INSIDE FACE OF PERIMETER INSULATION BOARD (TO PROTECT RESILIENT MATERIAL FROM MOISTURE)

PARTITIONS BORDERING FLOATED FLOORS SHOULD BE DESIGNED AS SHOWN IN "A" BELOW. IF PARTITIONS ARE SET ON A FLOATED FLOOR, EITHER A COMPRESSED RESILIENT PAD AS IN "B" OR A SHEAR FAILURE IN THE FLOATED SURFACE AS IN "C" WILL PROVIDE A DIRECT FLANKING PATH.

CAULKING SEAL

CAULKING SEAL AROUND PERIMETER

GLASS-FIBER PERIMETER INSULATION

(A) THIS

FLANKING SOUND PATHS

(B)

SOUND LEAKS

(C)

not THIS

171

NOISE CRITERIA FOR MECHANICAL SYSTEMS

Preferred noise criteria (PNC) data can be used as a means for specifying permissible steady background noises for various indoor functional activities. PNC data are given in octave-band frequencies on the following page. The noise from the mechanical system must *not* exceed the PNC rating appropriate for an activity. To achieve this goal, the designer must select quiet mechanical equipment, use internal duct linings and sound attenuating mufflers, and prevent transmission of vibrations through the building structure.

Type of Space (and listening requirements)	*Range of Preferred Noise Criteria (PNC)**
Concert halls, opera houses, recital halls, drama theaters, etc. (for excellent listening conditions)	Not to exceed 20
Bedrooms, sleeping quarters, hospitals, residences, apartments, hotels, motels, etc. (for sleeping, resting, relaxing)	20 to 30
Auditoriums, theaters, radio/TV studios, music practice rooms, large meeting rooms, audio-visual facilities, large conference rooms, executive offices, churches, courtrooms, chapels, etc. (for very good listening conditions)	20 to 30
Private or semiprivate offices, small conference rooms, classrooms, reading rooms, libraries, etc. (for good listening conditions)	30 to 35
Large offices, reception areas, retail shops and stores, cafeterias, restaurants, gymnasiums, etc. (for fair listening conditions)	35 to 40
Lobbies, corridors, laboratory work spaces, drafting and engineering rooms, general secretarial areas, maintenance shops (such as for electrical equipment), etc. (for moderately fair listening conditions)	40 to 45
Kitchens, laundries, school and industrial shops, garages, machinery spaces, computer equipment rooms, etc.	45 to 55

*The minimum PNC rating should be used for buildings where high quality listening environments are desired. Use the maximum rating only where economy or the physical situation makes it impractical to use the minimum rating.

PREFERRED NOISE CRITERIA SOUND PRESSURE LEVEL TABLE

The PNC data below can be used to define permissible octave-band frequency sound pressure levels in decibels for the various activities listed on the preceding page.

Sound Pressure Level (dB)

PNC	63 Hz	125 Hz	250 Hz	500 Hz	1000 Hz	2000 Hz	4000 Hz	8000 Hz
15	43	35	28	21	15	10	8	8
20	46	39	32	26	20	15	13	13
25	49	43	37	31	25	20	18	18
30	52	46	41	35	30	25	23	23
35	55	50	45	40	35	30	28	28
40	59	54	50	45	40	35	33	33
45	63	58	54	50	45	41	38	38
50	66	62	58	54	50	46	43	43
55	70	66	62	59	55	51	48	48

CHECKLIST FOR EFFECTIVE MECHANICAL SYSTEM NOISE CONTROL

Always specify the quietest equipment available. Often precautions can be taken to operate equipment at low noise levels (e.g., fan-coil units with low-medium-high motor speeds can be sized to operate on medium speed at the maximum cooling conditions).

Avoid attaching vibrating or noisy equipment directly to the structural surfaces in a building. In addition, treat the walls and ceiling of mechanical spaces with generous amounts of sound-absorbing materials. In large buildings the mechanical space is often the major noise source, having noise levels in the range of 75 to 105 decibels.

Layers of soft, resilient materials can be used under the wearing surface of floors to isolate the mechanical equipment room from the structure or between the bases or supports of vibrating equipment and the structure to minimize transfer of vibrations into the structure. Use wall and floor-ceiling constructions with high sound transmission class ratings.

Prefabricated, resilient equipment-mounts and bases also are commercially available and, when properly selected, provide stable support to vibrating equipment.

Pipe and conduit connections to vibrating equipment should be vibration-isolated for a considerable distance to interrupt the potential transmission path. All electrical conduits and fluid pipe connections should be flexible and "floppy" when possible. Avoid metal-to-metal contact.

Avoid constructions such as untreated mechanical ducts or rigid conduits that can act as speaking tubes to transmit sound from one area to another. Line common ducts with glass-fiber and, where they pass through walls and floor-ceiling constructions, isolate them from the structure with resilient materials and caulk the perimeter airtight.

SELECTED REFERENCES

SELECTED REFERENCES

REFERENCES

1. Andrews, F. T., *The Architect's Guide to Mechanical Systems.* New York: Reinhold Publishing Corporation, 1966.

2. *ASHRAE Handbook & Product Directory* (formerly *ASHRAE Guide and Data Book)*, Four volumes: Applications, Fundamentals, Equipment, and Systems (published periodically).

3. Bruce, W., *Man and His Thermal Environment.* Ottawa, Canada: National Research Council, 1960.

4. Conklin, G., *The Weather Conditioned House.* New York: Reinhold Publishing Corporation, 1958.

5. Egan, M. D., *Concepts in Architectural Acoustics.* New York: McGraw-Hill, Inc., 1972.

6. Fischer, R. E. (ed.), *Architectural Engineering—Environmental Control.* New York: McGraw-Hill, Inc., 1965.

7. Fitch, J. M., *American Building 2: The Environmental Forces That Shape It.* Boston, Massachusetts: Houghton Mifflin Company, 1972.

8. Greenberg, A., "Heating, Ventilating, and Air Conditioning" in J. H. Callender (ed.), *Time-saver Standards.* New York: McGraw-Hill, Inc., 1966.

9. Kent, S. R. (ed.), *The Environmental Services of Buildings.* Toronto, Canada: Ontario Association of Architects, 1970.

10. Kinzey, B. Y. and H. M. Sharp, *Environmental Technologies in Architecture.* Englewood Cliffs, New Jersey: Prentice-Hall, Inc., 1963.

11. Laube, H. L., *How to Have Air-Conditioning and Still Be Comfortable.* Birmingham, Michigan: Business News Publishing Company, 1971.

12. McGuiness, W. J. and B. Stein, *Mechanical and Electrical Equipment for Buildings.* New York: John Wiley & Sons, Inc., 1971.

13. Olgyay, V., *Design with Climate.* Princeton, New Jersey: Princeton University Press, 1963.

14. Olivieri, J. B., *How to Design Heating-Cooling Comfort Systems.* Birmingham, Michigan: Business News Publishing Company, 1971.

15. Ramsey, C. G. and H. R. Sleeper, *Architectural Graphic Standards.* New York: John Wiley & Sons, Inc., 1971.

16. Reiner, L.E., *Methods and Materials of Construction.* Englewood Cliffs, New Jersey: Prentice-Hall, Inc., 1970.

17. Rogers, T. S., *Thermal Design of Buildings.* New York: John Wiley & Sons, Inc., 1964.

18. *Trane Air Conditioning Manual.* LaCrosse, Wisconsin: The Trane Company, 1969.

APPENDICES

APPENDIX A: SUMMARY OF USEFUL FORMULAS

General

Basic Thermal Comfort:

$$M \pm \Delta S = \pm C \pm R + E \qquad \text{(A-1)}$$

where M = metabolism (varies with activity and emotional state) in Btuh

ΔS = withdrawals from, or additions to, body heat storage in Btuh

C = convection heat loss or gain in Btuh

R = radiation heat loss or gain in Btuh

E = evaporation heat loss in Btuh

Note: The "+" sign indicates heat losses, the "–" sign heat gains.

Celsius (or Centigrade) Temperature:

$$t = 1.8t_c + 32 \qquad \text{(A-2)}$$

where t = temperature in degrees fahrenheit (°F)

t_c = temperature in degrees centigrade (°C)

Note: Convert temperature (t_c) in degrees centigrade to metric temperature in degrees kelvin (°K) by adding 273 to the t_c value.

Mean Radiant Temperature:

$$\text{MRT} = \frac{\Sigma \, At}{\Sigma \, A} \qquad \text{(A-3)}$$

or

$$\text{MRT} = \frac{A_1 t_1 + A_2 t_2 + \ldots + A_n t_n}{A_1 + A_2 + \ldots + A_n}$$

where   MRT = mean radiant temperature (i.e., the weighted average of the various radiant heat influences in a space) in °F

A = projected area of a specific surface or object in feet, radians, or degrees

t = surface temperature in °F

Climate And Shelter

Shade from Overhangs:

$$d = x \, (\tan \alpha / \cos \beta) \qquad \text{(A-4)}$$

where d = depth of shade in feet
   x = width of overhang in feet
   α = altitude angle in degrees
   β = bearing (or azimuth) angle in degrees

Natural Ventilation:

$$Q \simeq 50Av \qquad \text{(A-5)}$$

where Q = airflow volume in cubic feet per minute (cfm)
   A = open area of inlets (or outlets) in ft^2
   v = wind velocity in mph

Heat Flow

$$U = \frac{1}{R_T} \qquad \text{(A-6)}$$

where U = overall coefficient of heat transmission (called "U-value") in Btuh/ft^2/°F
   R_T = total thermal resistance in hrs/Btu/ft^2/°F

$$R_T = \left(\frac{1}{f_i} + \frac{x}{k} + \frac{1}{a} + \frac{1}{C} + \frac{1}{f_o}\right) \qquad \text{(A-7)}$$

where R_T = [See Equation (A-6)]

   f_i = inside air film conductance in Btuh/ft^2/°F

   x = thickness of homogenous material in inches
   k = conductivity in Btuh/ft^2/in./°F
   a = air space conductance in Btuh/ft^2/°F
   C = conductance in Btuh/ft^2/°F (conductances for building materials are specified for a given thickness)
   f_o = outside air film conductance in Btuh/ft^2/°F

Attic Spaces:

$$U = \frac{U_r U_c}{U_r + (U_c/n)} \qquad \text{(A-8)}$$

where U = U-value for an attic space in Btuh/ft^2/°F
   U_r = U-value of roof construction in Btuh/ft^2/°F
   U_c = U-value of ceiling construction in Btuh/ft^2/°F
   n = ratio of roof area to ceiling area (no units)

Dew-point Analysis:

$$\frac{R_T}{(t - t_o)} = \frac{R_i}{(t - t_d)} \qquad \text{(A-9)}$$

where R_T = [See Equation (A-6)]

t_o = outside air temperature in °F

t = inside air temperature in °F

R_i = inside air film resistance ($1/f_i$) in hrs/Btu/ft²/°F

t_d = dew-point temperature in °F (temperature that will cause condensation on inside glass surface when a specific room relative humidity value is exceeded)

Heat Loss

$$H_c = AU(t - t_o) \qquad \text{(A-10)}$$

where H_c = heat loss (conduction) in Btuh

A = area of wall, glass, or roof surface in ft²

U = [See Equation (A-6)]

t = inside air temperature in °F

t_o = outside air temperature in °F

$$H_e = FP(t - t_o) \qquad \text{(A-11)}$$

where H_e = heat loss (conduction) through floor slab edges in Btuh

F = factor of 0.81 for floors without edge insulation; 0.55 for floors with edge insulation

P = perimeter of slab in feet

$(t - t_o)$ = [See Equation (A-10)]

$$H_i = 0.018q(t - t_o) \qquad \text{(A-12)}$$

where H_i = heat loss (convection) in Btuh

q = infiltration airflow volume in cubic feet per hour (from tables)

$(t - t_o)$ = [See Equation (A-10)]

$$H_s = FA \qquad \text{(A-13)}$$

where H_s = heat loss (conduction) through floor slabs or basement walls in Btuh

F = factor of 2 for floor slabs; 4 for basement walls

A = floor slab (or basement wall) area in ft²

Heat Gain

$$H = AU \, (ETD) \qquad \text{(A-14)}$$

where H = heat gain (conduction and radiation) in Btuh

              A = area of opaque wall or roof surface in ft²

              U = [See Equation (A-6)]

    ETD = equivalent temperature differential in °F (used to account for thermal time-lag and solar radiation effects)

Glass:

$$H_c = AU \, (t_o - t) \qquad \text{(A-15)}$$

where H_c = heat gain (conduction) in Btuh

              A = area of glass surface in ft²

              U = [See Equation (A-6)]

   $(t_o - t)$ = [See Equation (A-10)]

$$H_o = AS_g \, (S.C.) \qquad \text{(A-16)}$$

where H_o = heat gain (radiation) for a given orientation in Btuh

           A = area of opening (wall or roof) in ft²

          S_g = solar heat gain factor in Btuh/ft²

   $S.C.$ = shading coefficient (no units)

Electrical and Mechanical Equipment:

$$H_m = 3.4 \, W \qquad \text{(A-17)}$$

where H_m = heat gain in Btuh (increase by 25% for fluorescent and other arc-type lights to account for heat generated in ballasts)

         W = lighting, electrical and mechanical equipment energy in watts

Note: Include sensible heat gains from building occupants in total used to size air-handling units and ducts.

Mechanical Systems

$$Q = \frac{H.L.}{1.08 \, (t_e - t)} \qquad \text{(A-18)}$$

where Q = heating airflow volume in cfm

     $H.L.$ = heat loss (room or building) in Btuh

        t_e = temperature at heating equipment (called "diffusion" temperature) in °F

        t = room or building air temperature in °F

$$Q = \frac{H_{sen}}{1.08\,(t - t_d)\,(1 - b)} \qquad \text{(A-19)}$$

where Q = cooling airflow volume in cfm

         H_{sen} = sensible heat gain (room or building) in Btuh

         t = [See Equation (A-18)]

         t_d = dew-point temperature of equipment in °F

         b = bypass factor of cooling coil (depends on the physical and operating characteristics of the equipment)

$$N = 60\,\frac{Q}{V} \qquad \text{(A-20)}$$

where N = number of air changes per hour (ach)

         Q = airflow volume in cfm

         V = volume (room or building) in ft³

Sensible Heat Ratio:

$$SHR = \frac{H_{sen}}{H_{sen} + H_{lat}} \qquad \text{(A-21)}$$

where SHR = sensible heat ratio (no units)

         H_{sen} = sensible heat gain in Btuh

         H_{lat} = latent heat gain in Btuh

Refrigeration Load:

$$H.G. = 4.5Q\,\Delta h \qquad \text{(A-22)}$$

where $H.G.$ = total heat load (sensible and latent) at cooling coils in Btuh

         Q = [See Equations (A-19) and (A-20)]

         Δh = enthalpy in Btu/lb

$$v = \frac{Q}{A} \qquad \text{(A-23)}$$

where v = air velocity in feet per minute (fpm)

         Q = airflow volume in cfm

         A = area (open or "free") of duct or grille in ft² (if area is in square inches multiply equation by 144)

$$d = \sqrt{1.3A} \qquad \text{(A-24)}$$

where d = round duct diameter in inches

       A = duct cross-section area in square inches

Note: To find round duct diameter (d) directly, use $d = 13.7\sqrt{Q/v}$

Indoor-outdoor Air Balance:

$$Q_S = Q_r + Q_O \qquad \text{(A-25)}$$

where Q_S = supply airflow volume in cfm

       Q_r = return airflow volume in cfm

       Q_O = outdoor (or "make-up") airflow volume in cfm

Duct Heat Loss in Unconditioned Spaces:

$$H_w \simeq 20Pl \qquad \text{(A-26)}$$

where H_w = heat loss through insulated duct in Btuh

      P = perimeter of duct in feet

      l = length of duct in feet

Fan Heat:

$$H_f \simeq \frac{2545\ H.P.}{E} \qquad \text{(A-27)}$$

where H_f = fan heat gain in Btuh

    $H.P.$ = fan horsepower rating

      E = motor efficiency (use 0.85 for large motors; 1.0 if fan is not located in conditioned space)

Mechanical System Noise and Vibrations

$$f_n = 3.13\sqrt{1/y} \qquad \text{(A-28)}$$

where f_n = natural frequency in Hz

     y = system static deflection in inches

$$\frac{\Omega}{f_n} \geqslant 3 \qquad \text{(A-29)}$$

where Ω = equipment forcing frequency in Hz

     f_n = [See Equation (A-28)]

Miscellaneous

Fuel Savings:

$$F = \frac{24\,(U - U_i)nA}{CE} \qquad\qquad \text{(A-30)}$$

where

F = fuel savings in watts, cu ft, or gal (depends on type of fuel)

$(U - U_i)$ = uninsulated less insulated U-value in Btuh/ft²/°F

n = degree days

A = surface area in ft²

CE = heat value and efficiency coefficient (use 3.4 for electricity, 800 for natural gas, and 100,000 for oil)

Spray Ponds:

$$S_a \simeq 16T \qquad\qquad \text{(A-31)}$$

where

S_a = spray pond surface area in ft²

T = refrigeration load in tons

APPENDIX B: DEW-POINT TABLE

Dry Bulb Temp. (°F)	RELATIVE HUMIDITY (%)									
	22	24	26	28	30	32	34	36	38	40
	Dew-point Temperature (°F)									
30	1.0	0.7	2.2	3.7	5.0	6.3	7.5	8.6	9.7	10.7
35	3.1	4.8	6.4	7.9	9.2	10.5	11.7	12.9	14.0	15.0
40	7.0	8.7	10.3	11.8	13.2	14.5	15.7	16.9	18.1	19.1
45	10.8	12.6	14.2	15.7	17.2	18.5	19.8	21.0	22.1	23.2
50	14.6	16.3	18.1	19.6	21.1	22.4	23.7	25.0	26.1	27.2
55	18.4	20.3	21.9	23.5	25.0	26.4	27.7	29.0	30.1	31.3
60	22.2	24.1	25.8	27.4	28.9	30.3	31.7	33.0	34.4	35.7
61	23.0	24.8	26.6	28.2	29.7	31.1	32.5	33.9	35.3	36.6
62	23.7	25.6	27.3	29.0	30.5	31.9	33.4	34.8	36.2	37.5
63	24.5	26.4	28.1	29.7	31.2	32.7	34.2	35.7	37.0	38.3
64	25.2	27.1	28.9	30.5	32.0	33.6	35.1	36.6	37.9	39.2
65	26.0	27.9	29.6	31.3	32.9	34.5	36.0	37.4	38.8	40.1
66	26.7	28.6	30.4	32.0	33.7	35.4	36.9	38.3	39.7	41.0
67	27.5	29.4	31.2	32.9	34.6	36.2	37.8	39.2	40.6	41.9
68	28.3	30.2	31.9	33.7	35.5	37.1	38.6	40.1	41.5	42.8
69	29.0	30.9	32.8	34.6	36.3	38.0	39.5	41.0	42.4	43.7
70	29.8	31.7	33.6	35.5	37.2	38.8	40.4	41.9	43.3	44.6
71	30.5	32.5	34.5	36.3	38.1	39.7	41.3	42.8	44.2	45.5
72	31.2	33.3	35.3	37.2	38.9	40.6	42.1	43.6	45.1	46.4
73	32.0	34.1	36.2	38.0	39.8	41.5	43.0	44.5	45.9	47.3
74	32.8	35.0	37.0	38.9	40.7	42.3	43.9	45.4	46.8	48.2
75	33.7	35.8	37.9	39.7	41.5	43.2	44.8	46.3	47.7	49.1
76	34.5	36.7	38.7	40.6	42.4	44.1	45.7	47.2	48.6	50.0
77	35.3	37.5	39.6	41.5	43.2	44.9	46.5	48.0	49.5	50.9
78	36.1	38.3	40.4	42.3	44.1	45.8	47.4	48.9	50.4	51.8
79	37.0	39.2	41.2	43.2	45.0	46.7	48.3	49.8	51.3	52.7
80	37.8	40.0	42.1	44.0	45.8	47.5	49.2	50.7	52.2	53.5
81	38.6	40.9	42.9	44.9	46.7	48.4	50.0	51.6	53.0	54.4
82	39.5	41.7	43.8	45.7	47.6	49.3	50.9	52.4	53.9	55.3
83	40.3	42.5	44.6	46.6	48.4	50.1	51.8	53.3	54.8	56.2
84	41.1	43.4	45.5	47.4	49.3	51.0	52.7	54.2	55.7	57.1
85	42.0	44.2	46.3	48.3	50.1	51.9	53.5	55.1	56.6	58.0
86	42.8	45.1	47.2	49.1	51.0	52.7	54.4	56.0	57.5	58.9
87	43.6	45.9	48.0	50.0	51.9	53.6	55.3	56.9	58.3	59.8
88	44.4	46.7	48.9	50.8	52.7	54.5	56.1	57.7	59.2	60.7
89	45.3	47.6	49.7	51.7	53.6	55.3	57.0	58.6	60.1	61.6
90	46.1	48.4	50.5	52.6	54.4	56.2	57.9	59.5	61.0	62.5
92	47.8	50.1	52.2	54.3	56.2	57.9	59.6	61.2	62.8	64.2
94	49.4	51.7	53.9	56.0	57.9	59.7	61.4	63.0	64.5	66.0
96	51.1	53.4	55.6	57.7	59.6	61.4	63.1	64.7	66.3	67.8
98	52.7	55.1	57.3	59.4	61.3	63.1	64.9	66.5	68.1	69.6
100	54.4	56.7	59.0	61.1	63.0	64.9	66.6	68.2	69.8	71.3
102	56.0	58.4	60.7	62.7	64.7	66.6	68.3	70.0	71.6	73.1
104	57.6	60.1	62.3	64.4	66.4	68.3	70.1	71.7	73.3	74.9
106	59.3	61.7	64.0	66.1	68.1	70.0	71.8	73.5	75.1	76.6
108	60.9	63.4	65.7	67.8	69.8	71.7	73.5	75.2	76.9	78.4
110	62.6	65.1	67.4	69.5	71.5	73.5	75.3	77.0	78.6	80.2

Dry Bulb Temp. (°F)	RELATIVE HUMIDITY (%)									
	42	44	46	48	50	52	54	56	58	60
				Dew-point Temperature (°F)						
30	11.7	12.7	13.6	14.4	15.3	16.1	16.9	17.6	18.3	19.1
35	16.0	17.0	17.9	18.8	19.7	20.5	21.3	22.1	22.8	23.5
40	20.1	21.1	22.1	23.0	23.8	24.7	25.5	26.3	27.0	27.8
45	24.2	25.2	26.1	27.1	28.0	28.8	29.6	30.4	31.2	32.0
50	28.3	29.3	30.3	31.2	32.1	33.1	34.1	35.0	35.8	36.7
55	32.4	33.6	34.7	35.7	36.7	37.7	38.6	39.6	40.5	41.4
60	36.9	38.1	39.2	40.3	41.3	42.3	43.3	44.3	45.2	46.1
61	37.8	39.0	40.1	41.2	42.2	43.3	44.3	45.2	46.2	47.1
62	38.7	39.9	41.0	42.1	43.2	44.2	45.2	46.2	47.1	48.0
63	39.5	40.7	41.9	43.1	44.1	45.1	46.1	47.1	48.0	48.9
64	40.5	41.7	42.9	44.0	45.1	46.1	47.1	48.0	49.0	49.9
65	41.4	42.6	43.8	44.9	45.9	47.0	48.0	49.0	49.9	50.8
66	42.3	43.5	44.6	45.7	46.8	47.9	48.9	49.9	50.8	51.8
67	43.2	44.4	45.6	46.7	47.8	48.8	49.8	50.8	51.8	52.7
68	44.1	45.3	46.5	47.6	48.7	49.7	50.8	51.8	52.7	53.7
69	45.0	46.2	47.4	48.5	49.6	50.7	51.7	52.7	53.7	54.6
70	45.9	47.2	48.3	49.4	50.5	51.6	52.6	53.6	54.6	55.5
71	46.8	48.0	49.2	50.3	51.4	52.5	53.5	54.5	55.5	56.4
72	47.7	48.9	50.1	51.3	52.4	53.5	54.5	55.4	56.4	57.4
73	48.6	49.8	51.0	52.2	53.3	54.4	55.4	56.4	57.4	58.3
74	49.5	50.8	52.0	53.1	54.2	55.3	56.3	57.3	58.3	59.2
75	50.4	51.6	52.9	54.0	55.1	56.2	57.3	58.3	59.2	60.2
76	51.3	52.5	53.8	54.9	56.0	57.1	58.2	59.2	60.2	61.1
77	52.2	53.5	54.7	55.8	57.0	58.1	59.1	60.1	61.1	62.1
78	53.1	54.4	55.6	56.7	57.9	59.0	60.1	61.1	62.0	63.0
79	54.0	55.3	56.5	57.7	58.8	59.9	61.0	62.0	63.0	63.9
80	54.9	56.2	57.4	58.6	59.7	60.8	61.9	62.9	63.9	64.9
81	55.8	57.1	58.3	59.5	60.6	61.7	62.8	63.8	64.8	65.8
82	56.7	58.0	59.2	60.4	61.5	62.7	63.7	64.8	65.8	66.8
83	57.5	58.8	60.1	61.3	62.5	63.6	64.6	65.7	66.7	67.7
84	58.5	59.8	61.0	62.2	63.4	64.5	65.6	66.6	67.6	68.6
85	59.3	60.7	61.9	63.1	64.3	65.4	66.5	67.5	68.6	69.6
86	60.3	61.6	62.8	64.0	65.2	66.3	67.4	68.5	69.5	70.5
87	61.2	62.5	63.7	65.0	66.1	67.2	68.3	69.4	70.4	71.4
88	62.0	63.4	64.6	65.9	67.0	68.2	69.3	70.3	71.4	72.4
89	62.9	64.3	65.6	66.8	67.9	69.1	70.2	71.3	72.3	73.3
90	63.9	65.2	66.5	67.7	68.9	70.0	71.1	72.2	73.2	74.2
92	65.6	67.0	68.3	69.5	70.7	71.9	73.0	74.1	75.1	76.1
94	67.4	68.8	70.1	71.3	72.5	73.6	74.7	75.9	77.0	78.0
96	69.2	70.6	71.9	73.1	74.3	75.5	76.7	77.8	78.8	79.9
98	71.0	72.4	73.7	75.0	76.2	77.3	78.5	79.6	80.7	81.7
100	72.8	74.2	75.5	76.8	78.0	79.2	80.3	81.5	82.6	83.6
102	74.6	75.9	77.3	78.6	79.8	81.0	82.2	83.3	84.4	85.5
104	76.3	77.8	79.1	80.4	81.6	82.9	84.0	85.2	86.3	87.3
106	78.1	79.5	80.9	82.2	83.5	84.7	85.9	87.0	88.1	89.2
108	79.9	81.3	82.7	84.0	85.3	86.5	87.7	88.9	90.0	91.1
110	81.7	83.1	84.5	85.8	87.1	88.4	89.6	90.7	91.8	92.9

APPENDIX C: HEATING AND COOLING LOADS

Heating and cooling loads must be verified by precise computations at the actual design conditions. The values given in the table are for preliminary estimates only and should *not* be used for final design purposes.

Type of Space	Heating Load* (Btuh/sq ft)	Cooling Load** (sq ft/ton)	Air Quantity*** (cfm/sq ft)
Bedrooms, sleeping quarters, residences, hotels, motels, apartments, etc.	20 to 55	450 to 250	0.5 to 1.5
Offices, banks, schools, libraries, retail shops and stores, museums, etc.	25 to 65	350 to 150	1.0 to 2.5
Restaurants, cafeterias, laboratory work spaces, school and industrial shops, etc.	15 to 45	150 to 80	2.0 to 3.0
Auditoriums, large meeting rooms, theaters, large conference rooms, etc.	10 to 30	400 to 90	2.0 to 4.0

*Use upper end of the heating load range for locations having outdoor design temperatures less than 10°F and for buildings with glass areas greater than 40% of gross wall area.

**Use lower end of the cooling load range for buildings without indoor shading devices or with large glass areas.

***Air quantity range is based on cooling demand for all-air systems. It may be exceeded by ventilation requirements, e.g., where heavy smoking is anticipated.

APPENDIX D—FRICTION LOSS FOR ROUND ELBOWS (90°)

The curves below can be used to find the pressure loss in inches of water (''w.g.) at round duct elbows for various air velocities (v). Add the pressure loss from the graph to the duct friction computed for the length between the intersection of duct centerlines.

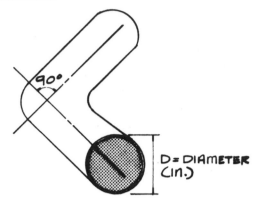

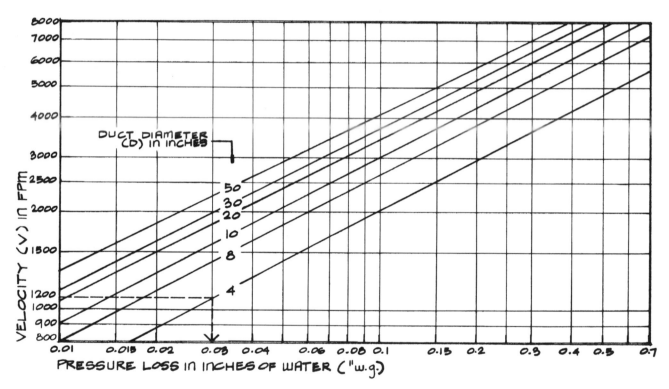

Example—Use of Graph

Given: Air velocity (v) of 1200 fpm at 4 in. elbow.

Procedure to find friction loss: Enter graph at v = 1200 fpm and read opposite D = 4 in. curve to pressure loss of 0.03 ''w.g. (See dashed lines on graph.)

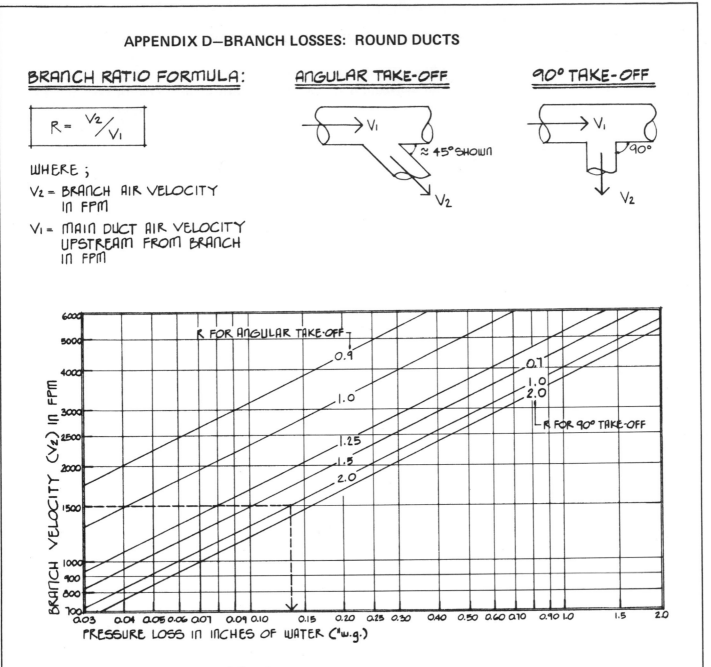

BRANCH RATIO FORMULA:

$$R = \frac{V_2}{V_1}$$

WHERE;

V_2 = BRANCH AIR VELOCITY IN FPM

V_1 = MAIN DUCT AIR VELOCITY UPSTREAM FROM BRANCH IN FPM

ANGULAR TAKE-OFF

90° TAKE-OFF

Example—Use of Graph

Given: Main duct air velocity (v_1) upstream from angular take-off is 750 fpm and branch velocity (v_2) is 1500 fpm.

Procedure to find branch pressure loss:

1. Calculate $R = v_2/v_1 = 1500/750 = 2.0$.
2. Enter graph at v_2 = 1500 fpm and read opposite R = 2.0 curve (for angular take-off) to pressure loss of 0.014 ''w.g. (See dashed lines on graph.)

Note: For comprehensive data on pressure losses at round duct fittings and components, see Eschman, R. and W.E. Long, "A Critical Assessment of High Velocity Duct Design Information." *ASHRAE Transactions,* Vol. 76, 1970.

APPENDIX E: CONVERSION FACTORS

The table below presents conversion factors for common thermal comfort and mechanical system units to corresponding metric system units, often referred to as the Système International d'Unités (SI units). The basic SI units are expressed as follows: length by meter (abbreviated: m), mass by kilogram (kg), time by second (s), and temperature by degree kelvin (K). For a comprehensive presentation of metric system units, refer to "Metric Practice Guide," ASTM Designation: E 380-72. This publication is available from the American Society for Testing and Materials (ASTM), 1916 Race Street, Philadelphia, PA 19103.

To Convert	Into	Multiply by*	Conversely, Multiply by*
Btu	joule	1055	9.4845×10^{-4}
Btuh	watt	0.293	3.413
Btuh/ft^2	watt/m^2	2.722×10^{-2}	36.73
Btuh/ft^2/$^\circ$F	watt/m^2/$^\circ$K	5.678	0.1763
Btuh/ft^2/in./$^\circ$F	watt/m/$^\circ$K	1.442×10^{-1}	6.934
Btu/lb	calorie/g	0.5556	1.8
	joule/kg	2.326×10^3	4.299×10^{-4}
$^\circ$C	$^\circ$F	($^\circ$C $\times$ 9/5) + 32	($^\circ$F−32) $\times$ 5/9
	$^\circ$K	$^\circ$C + 273	$^\circ$K − 273
cm	in.	0.3937	2.54
	ft	3.281×10^{-2}	30.48
	mm	10	10^{-1}
	m	10^{-2}	10^2
cfh	m^3/s	7.865×10^{-6}	1.271×10^5
cfm	m^3/s	4.719×10^{-4}	2119
deg (angle)	radian	1.745×10^{-2}	57.3
$^\circ$F	$^\circ$C	($^\circ$F − 32) $\times$ 5/9	($^\circ$C $\times$ 9/5) + 32
	$^\circ$K	($^\circ$F + 460) $\times$ 5/9	1.8°K − 460
ft	in.	12	0.0833
	mm	304.8	3.281×10^{-3}
	cm	30.48	3.281×10^{-2}
	m	0.3048	3.281
ft^2	in.^{2}	144	6.945×10^{-3}
	cm^2	9.290×10^2	0.01076
	m^2	9.290×10^{-2}	10.76
ft^3	in.^{3}	1728	5.787×10^{-4}
	cm^3	2.832×10^4	3.531×10^{-5}
	m^3	2.832×10^{-2}	35.31
fpm	m/s	5.08×10^{-3}	197
	mph	1.136×10^{-2}	88
gpm	m^3/s	6.309×10^{-5}	1.585×10^4
in.	ft	0.0833	12
	mm	25.4	0.03937
	cm	2.54	0.3937
	m	0.0254	39.37
"w.g.	pascal	2.4884×10^2	4.019×10^{-3}

*Round converted quantity to the proper number of significant digits commensurate with its intended precision.

APPENDIX E: CONVERSION FACTORS (Continued)

To Convert	Into	Multiply by*	Conversely, Multiply by*
°K	°C	°K − 273	°C + 273
	°F	1.8°K − 460	(°F + 460) X 5/9
	°R (Rankine)	1.8°K	°R X 5/9
m	in.	39.37	0.0254
	ft	3.281	0.3048
	yd	1.0936	0.9144
	mm	10^3	10^{-3}
	cm	10^2	10^{-2}
m/s	fpm	197	5.08×10^{-3}
miles	ft	5280	1.894×10^{-4}
	km	1.6093	0.6214
mph	fpm	88	1.136×10^{-2}
	km/min	2.682×10^{-2}	37.28
	km/hr	1.6093	0.6214
lbs (weight)	gm	453.6	2.205×10^{-3}
	kg	0.4536	2.205
ton (refrigeration)	kilowatt	3.517	0.2843
	watt	3517	2.843×10^{-4}
watt	horsepower (hp)	1.341×10^{-3}	745.7

*Round converted quantity to the proper number of significant digits commensurate with its intended precision.

INDEX[*]

* For easy reference for checklists, example problems, and tables—see pages 201, 202, 203.

K

Kansas State University, 9
Kent, S. R., 176
Kinzey, B. Y., 176
Koestel, A., 11

L

latent heat, 81
lateral air duct system, 111
Laube, H. L., 176
Long, W. E., 189

M

main ducts, 98, 100, 104
make-up air, 135, 140
McGuiness, W. J., 176
mean radiant temperature (MRT)
    (*See* temperature)
mechanical duct system noise
    isolation, 162, 166–68
mechanical equipment rooms, 114,
    162, 163
metabolism, 4
metric system units, 190, 191
mixing box, 100
moisture:
    barriers, 58
    content of air, 17
    problems in winter months, 9
    sources in buildings, 16
mufflers, 162, 168

N

National Climate Center, 80
National Fire Protection
    Association, 119
National Warm Air Heating & Air
    Conditioning Association,
    14

National Weather Service, 40, 72,
    75, 80, 87
natural frequency, 159, 160
natural ventilation, 38
Nevins, R. G., 9
noise criteria, 172

O

odor control (*See* outdoor air)
Olgyay, V., 20, 22, 25, 28, 30, 39,
    176
Olivieri, J. B., 108, 176
opposed-blade dampers, 126
orientations, building (*See* building
    shapes)
outdoor air, 119, 135, 140
overall coefficient of heat
    transmission (*See* *U*-value)
oxygen requirements, 15

P

packaged terminal air conditioner
    (PTAC), 94, 106
Park, K. S., 106
Penwarden, A. D., 39
perimeter loop air duct system, 111
perm, 57, 59
permeance, 59
psychrometric chart, 138

R

radial air duct system, 111
radiant panel systems, 84, 94, 105
radiation:
    basic factors, 2
    definition, 47
    effects, 6
    germicidal action, 37
radius ratio, 124, 126
Ramsey, C. G., 33, 176
reciprocals, 65
refrigeration load, 135–37

CHECKLISTS

EXAMPLE PROBLEMS

TABLES